SpringerBriefs in Energy

SpringerBriefs in Energy presents concise summaries of cutting-edge research and practical applications in all aspects of Energy. Featuring compact volumes of 50 to 125 pages, the series covers a range of content from professional to academic. Typical topics might include:

- A snapshot of a hot or emerging topic
- A contextual literature review
- A timely report of state-of-the art analytical techniques
- An in-depth case study
- A presentation of core concepts that students must understand in order to make independent contributions.

Briefs allow authors to present their ideas and readers to absorb them with minimal time investment.

Briefs will be published as part of Springer's eBook collection, with millions of users worldwide. In addition, Briefs will be available for individual print and electronic purchase. Briefs are characterized by fast, global electronic dissemination, standard publishing contracts, easy-to-use manuscript preparation and formatting guidelines, and expedited production schedules. We aim for publication 8–12 weeks after acceptance.

Both solicited and unsolicited manuscripts are considered for publication in this series. Briefs can also arise from the scale up of a planned chapter. Instead of simply contributing to an edited volume, the author gets an authored book with the space necessary to provide more data, fundamentals and background on the subject, methodology, future outlook, etc.

SpringerBriefs in Energy contains a distinct subseries focusing on Energy Analysis and edited by Charles Hall, State University of New York. Books for this subseries will emphasize quantitative accounting of energy use and availability, including the potential and limitations of new technologies in terms of energy returned on energy invested. The second distinct subseries connected to SpringerBriefs in Energy, entitled Computational Modeling of Energy Systems, is edited by Thomas Nagel, and Haibing Shao, Helmholtz Centre for Environmental Research - UFZ, Leipzig, Germany. This sub-series publishes titles focusing on the role that computer-aided engineering (CAE) plays in advancing various engineering sectors, particularly in the context of transforming energy systems towards renewable sources, decentralized landscapes, and smart grids.

All Springer brief titles should undergo standard single-blind peer-review to ensure high scientific quality by at least two experts in the field.

Aurelia Rybak · Aleksandra Rybak

The Role of Clean Coal Technologies in Energy Transformation and Energy Security

Ensuring Energy Security: The Key Role of Clean Coal Technologies in the Energy Transition

Aurelia Rybak
Department of Electrical Engineering
and Industrial Automation
Faculty of Mining, Safety Engineering
and Industrial Automation
Silesian University of Technology
Gliwice, Poland

Aleksandra Rybak
Department of Physical Chemistry
and Technology of Polymers
Faculty of Chemistry
Silesian University of Technology
Gliwice, Poland

ISSN 2191-5520 ISSN 2191-5539 (electronic)
SpringerBriefs in Energy
ISBN 978-3-031-80651-3 ISBN 978-3-031-80652-0 (eBook)
https://doi.org/10.1007/978-3-031-80652-0

© The Editor(s) (if applicable) and The Author(s), under exclusive license to Springer Nature Switzerland AG 2025

This work is subject to copyright. All rights are solely and exclusively licensed by the Publisher, whether the whole or part of the material is concerned, specifically the rights of translation, reprinting, reuse of illustrations, recitation, broadcasting, reproduction on microfilms or in any other physical way, and transmission or information storage and retrieval, electronic adaptation, computer software, or by similar or dissimilar methodology now known or hereafter developed.
The use of general descriptive names, registered names, trademarks, service marks, etc. in this publication does not imply, even in the absence of a specific statement, that such names are exempt from the relevant protective laws and regulations and therefore free for general use.
The publisher, the authors and the editors are safe to assume that the advice and information in this book are believed to be true and accurate at the date of publication. Neither the publisher nor the authors or the editors give a warranty, expressed or implied, with respect to the material contained herein or for any errors or omissions that may have been made. The publisher remains neutral with regard to jurisdictional claims in published maps and institutional affiliations.

This Springer imprint is published by the registered company Springer Nature Switzerland AG
The registered company address is: Gewerbestrasse 11, 6330 Cham, Switzerland

If disposing of this product, please recycle the paper.

Preface

The presented book discusses issues related to clean coal technologies and their role in the energy transition. The analyses carried out concern the most important countries still relying on coal, i.e., China, India, Australia, and the USA, with particular emphasis on the European Union. Poland was considered the most representative country in this case, due to the available coal resources in its geographical area, as well as due to the large share of coal in the structure of energy production in this country. Additionally, Poland, as one of the EU member states, has an exceptionally complicated task—the energy transformation and decarbonisation of the energy system by 2050. The restrictive standards and expectations of the EU towards its members put Poland in an unprecedented position on a global scale. Either Poland will apply CCT and renewable energy and turn out to be a great winner, or it will suffer a failure, the main consequences of which will fall on its citizens. If the implementation of CCT is successful in Poland, there is a great probability that it will be successful in every country in the world.

The study focusses on coal because clean coal technologies have been created for coal and it is mainly coal that is the subject of most global decarbonisation treaties. However, other fossil fuels also generate the same harmful substances during combustion. In the case of crude oil, CO_2 emissions are only 20% lower, and for natural gas by about 50%. In the near future, these energy carriers will also have to disappear from the energy generation structure. Therefore, it will be necessary to develop technology that allows the use of fossil fuels in an ecological way, and the sooner actions are taken to build them, the better it will affect the stability in energy systems of countries around the world.

Gliwice, Poland

Aurelia Rybak
Aleksandra Rybak

Contents

1 **Introduction** 1
   1.1   Definition of Clean Coal Technologies 4
   1.2   The Role of Clean Technologies in the Energy Transition 6
   1.3   Challenges and Benefits of Using Clean Coal Technologies 8
   References 9

2 **Coal Market** 13
   2.1   Market Forecasts for the Coal Sector 15
   2.2   Alternative Energy Sources 24
   References 25

3 **Clean Coal Technologies (CCT)** 27
   3.1   Membrane Techniques 30
   3.2   Waste Management from the Coal Production and Combustion Process and Its Role in the Development of Renewable Energy Sources 32
      3.2.1   Solid Waste 32
      3.2.2   Gaseous Waste 34
   References 37

4 **Coal Energy Technologies and Renewable Energy Sources** 41
   4.1   Hybrid Energy Solutions 41
   4.2   Energy-Chemical Clusters Profitability Analysis 45
   References 47

5 **Clean Coal Technologies and Energy Security** 49
   References 53

6	**Prospects for the Development of Clean Coal Technologies**	55
	6.1 New Directions of Research in the Field of Clean Coal Technologies	56
	6.2 Challenges and Opportunities for Clean Coal Technologies	68
	6.3 The Future of Clean Coal Technologies	69
	References	70
7	**Summary**	75

Chapter 1
Introduction

Abstract This chapter presents two possible strategies for decarbonizing energy systems. Clean coal technologies are characterized and related concepts are defined. It also describes the role that clean coal technologies can play in the energy transformation process and how they can complement the energy system based on renewable energy sources.

Keywords Clean coal technologies · Energy transition · Rare earth elements

At the beginning of the twentieth century, the basic fuel used for the production of heat, electricity, and in transport was coal. It had been mined for thousands of years in China (3490 BC) [1], USA, Australia, India, Indonesia, and in Europe it was used mainly in Great Britain, Germany, Poland, the Czech Republic, Belgium, and Spain [2–4]. Its importance increased in the eighteenth century. Coal was the basis of the industrial revolution, economic development, energy, and military security for many decades. Access to energy also contributed to improved living conditions, access to education and health care [5]. Therefore, coal played an invaluable role in the development of among others Europe, the United States, China, especially in the nineteenth and twentieth centuries, influencing their economy, geopolitical situation, society, and shaped these today's modern states. However, its use had an impact on another important area, namely the natural environment. Therefore, after centuries of domination in energy mixes, the time has come for energy transformation. Coal began to be perceived as an energy carrier with a harmful impact on the environment and a number of steps were taken to reduce and, in the future, completely abandon the use of this fossil fuel. Figure 1.1 presents the share of coal in primary energy consumption in selected countries of the European Union in 1965 and 2022. Countries with a consistently high level of coal consumption and those with the lowest share were selected for comparison. In the case of all countries, the decrease in the share of coal in the selected period of time is very clear, in the Czech Republic it is 53%, in Poland 42%, in the case of Portugal the share of coal fell by 14%, and in 2022 coal was not used in this country for energy production.

© The Author(s), under exclusive license to Springer Nature Switzerland AG 2025
A. Rybak, *The Role of Clean Coal Technologies in Energy Transformation and Energy Security*, SpringerBriefs in Energy, https://doi.org/10.1007/978-3-031-80652-0_1

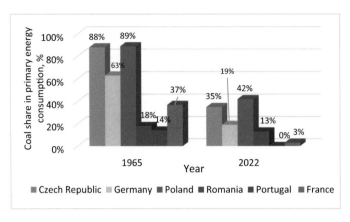

Fig. 1.1 Share of coal in primary energy consumption in 1965 and 2022 in Europe, own study based on statistical data from [8]

The complete removal of coal from the mixes of many countries will be an extremely complicated and expensive undertaking, paid for by numerous sacrifices on the part of EU citizens. At the same time, there are other countries rely heavily on coal. Currently, coal is a source of more than 50% of primary energy consumed in China and India [6]. They still intend to use this fuel using clean coal technologies, and in the perspective of 2050 its share in their mixes may reach even 35% [7]. Considering that China is responsible for 55% of global coal consumption, these forecasts will have a huge impact on the economy and the structure of coal fuel consumption on a global scale. For comparison, coal consumption in the European Union is less than 5% of global consumption.

As the demand for energy is constantly growing, other energy carriers and sources must take over the role of coal. In France, this is mainly nuclear energy, and in Germany and Poland, natural gas and oil. According to the guidelines of the Paris Agreement, SDG goals, the European Green Deal, and RED (Renewable Energy Directive) directives [10], by 2050, the EU Member States, and the USA are to decarbonise their energy systems and achieve climate neutrality. In China, this is 2060, India 2070 [11].

There are two ways to achieve this task. The first solution, which is promoted by the EU and implemented by most EU countries, is to completely abandon the use of coal and replace this source with renewable energy sources (RES). This solution is particularly difficult to implement in countries that still rely heavily on coal, especially when it comes to electricity generation. In this case, there are 10 EU countries that still rely heavily on coal, as shown in Fig. 1.2.

Coal is also the main source of electricity in China, India, and Australia (Fig. 1.3). Also, in these countries, efforts have been made to meet the growing demand for energy using alternative energy sources such as renewable energy, nuclear energy, or natural gas. As a result of these changes, the share of coal in electricity production

1 Introduction

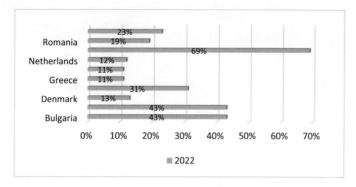

Fig. 1.2 Share of coal in electricity production in 2022 in selected EU countries, own study based on statistical data from [8]

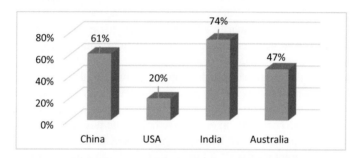

Fig. 1.3 Share of coal in electricity production in 2022, own study based on statistical data from [6]

in the USA dropped from 50% at the beginning of the twentieth century to around 20%.

As a result of recent events in the immediate vicinity of the European Union, namely the war in Ukraine, some countries have once again turned to coal, which has replaced the shortages in the supply of natural gas from the Russian Federation. Therefore, the share of coal in the mixes of countries such as Greece, the Czech Republic, Germany, Croatia, and Bulgaria has slightly increased.

The second solution that can be implemented on the path to decarbonisation is the use of clean coal technologies (CCT). It is particularly beneficial for coal-based countries, not only in the EU but all over the world. Clean coal technologies provide the possibility of clean coal combustion, which is very important in light of the growing demand for energy, as well as the need to transform energy systems. CCT would enable the stability of energy systems, ensuring energy security, while implementing the provisions of the Paris Agreement [11], the UN Sustainable Development Goals (SDGs) [12] and the European Green Deal [13]. The United Nations sees CCT as one of the pillars of the strategy to implement the Sustainable Development Goals, especially the Goal 7—Affordable and Clean Energy.

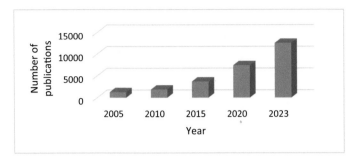

Fig. 1.4 Number of publications related to clean coal technologies according to ScienceDirect, own study based on statistical data from [14]

Clean Coal Technologies are attracting increasing interest worldwide. The number of publications related to this topic increased tenfold between 2005 and 2023 (Fig. 1.4) [14], which was primarily generated by countries where coal constitutes a significant percentage of the energy mix, such as China, India, the USA, Australia and Russia [15].

Unfortunately, due to the chemical composition of coal, the emission of carbon dioxide (CO_2), sulphur oxides (SO_x), nitrogen oxides (NO_x) and suspended particulate matter is unavoidable [16]. They are one of the causes of the greenhouse effect and global warming. Therefore, in order to maintain the possibility of burning coal and do it in a zero-emission way, it is necessary to use CCT.

Clean coal technologies are all solutions that allow the reduction of greenhouse gas emissions and other equally harmful chemical substances produced during coal combustion. Over the last 20 years, many different definitions and solutions have been developed that can be applied at every stage of the coal distribution process and the energy generation process. The most important are presented in the following chapters.

1.1 Definition of Clean Coal Technologies

Clean coal technologies (CCT) are technological solutions that allow its use in an environmentally neutral and economically justified manner [17]. They include solutions that are used to reduce the negative impact of coal extraction and combustion on the natural environment, and consequently climate change [18]. CCT is a technology designed to improve the efficiency of coal extraction, processing, transformation, and utilisation and to increase the acceptability of these processes from the point of view of the impact on the natural environment [19].

Clean coal technologies used around the world are in different stages of development. Some of them are in the research and implementation phase, others are fully

1.1 Definition of Clean Coal Technologies

researched and developed technologies that have been used for many years [20]. Generally, clean coal technologies can be divided into three categories:

- Used at the stage of coal enrichment—coal processing carried out at the stage of its exploitation, which allows for the removal of substances polluting the combustion process and increasing the process of greenhouse gas emissions. Due to its use, non-combustible parts are removed from the mined material, primarily in the form of gangue [21]. In this case, various technologies and methods, mechanical and chemical, can be used, such as jigs, heavy liquid separators, cyclones, spiral enrichers, selective grinding, deep enrichment, flotation machines [22–24], as well as biological methods using fungi and bacteria, the presence of which affects the sulphur and ash content in coal [25].
- Used during coal processing—there is a wide range of technologies that can be used to clean exhaust gases generated during coal combustion. These include primarily Carbon Capture and Storage CCS, Carbon Capture and Utilization CCU, Carbon Capture and Utilization and Sequestration [26, 27], fluidized bed boiler power plants [28], coal gasification in the bed and above ground gasification [29, 30] and membrane techniques [31].
- Used at the stage of management of waste generated during coal processing [32], such as gangue, water obtained during coal mining and processing, or CO_2 and methane [33, 34], as well as waste from the coal combustion process such as slag, mill waste, fly ash, CO_2, SO_x, NO_x.

The most beneficial way to implement CCT is to use the four-step clean coal strategy (CCS) taking into account the essence of R&D research on CCT, which will allow the refinement of existing solutions and the development of new ones (Fig. 1.5). The strategy takes into account and emphasises the role of CCT in the development of renewable energy sources, primarily due to the recovery of valuable substances and materials from waste necessary for the development of RES.

Fig. 1.5 Four-step strategy to implement the clean coal strategy (CCS), *source* own elaboration

1.2 The Role of Clean Technologies in the Energy Transition

The transformation of energy systems according to international treaties and guidelines consists of eliminating the use of fossil fuels such as coal, oil, natural gas and replacing them with renewable energy sources. This task is easier for some countries, especially Norway, Sweden, Switzerland, Canada, Brazil, or New Zealand, mainly due to the possibility of using water energy. The share of renewable energy sources in the total consumption of primary energy in these countries is currently in the range of 30–70%. They are in a privileged position due to their geographical location. In turn, countries whose geographical location did not provide them with such opportunities still rely largely on fossil fuels, including coal. It is of great importance primarily in countries that have access to domestic hard coal deposits. In the European Union, this is primarily Poland and Germany, whose resources amount to 3% of the world's proven reserves, as well as China, India, the USA, and Australia. Therefore, the natural order of things was to base the energy system of these countries on coal, which was also of great importance due to energy security in terms of energy availability and price. Renewable energy sources in the case of Poland developed very slowly until 2004, and the development was accelerated by accession to the EU. It was similar in the case of Hungary, Romania, the Czech Republic, and Bulgaria, which joined the EU in the fifth and sixth enlargements. The share of RES in primary energy consumption in Poland, the Czech Republic, and Hungary in 2022 was less than 10%. Despite being low, the share increases dynamically every year, considering that in 2007 it was close to zero [35]. In 2022, the share of RES in China, Australia was about 30%, in the USA and India 20%.

However, in order for RES to take over the role of coal, and in the future also the place of oil and natural gas, it will be necessary to accelerate and intensify the activities leading to their development. Implementing decarbonisation in Poland and in countries with similar specifics will require a special amount of work, as well as investment due to the composition of their energy mix. There are several main factors that influence the slow development of RES:

- Problems with ensuring stable and continuous operation of the energy system.
- Access to raw materials necessary to produce renewable energy technologies.
- No possibility of storing energy obtained from photovoltaics and wind turbines.
- The need to provide the financial resources necessary for the development of renewable energy.

For each of these, coal used in combination with clean coal technologies may provide a solution, as briefly described below.

The use of CCT will allow the production of energy necessary to maintain uninterrupted operation of the power grid. In addition, they will enable the recovery of valuable substances from waste of the fuel combustion process, such as gaseous waste and solid waste. The obtained CO_2 will find many applications, in the exploitation

1.2 The Role of Clean Technologies in the Energy Transition

of crude oil or food production. However, the most important is its use for the development of renewable energy sources, especially biomass [36], geothermal energy [37, 38] and energy storage. Energy storage will be the basis for energy systems based on renewable energy sources. Currently, these are mainly batteries that accumulate excess wind and solar energy, but alternative solutions are being developed for systems that compress industrial CO_2 into liquefied form and recover energy by expanding it at the time of energy demand. CO_2 in this case is used in a closed cycle, so it is not released into the atmosphere. This type of storage does not require access to expensive and strategic materials, and its location does not have to meet rigorous conditions [39].

Renewable energy sources require access to critical and strategic raw materials. This mainly concerns wind turbines, the development of which is impossible without access to rare earth elements such as Nd, Dy, Pr and Tb, as well as the solar energy, the effective use of which is out of the question without the application of energy storage. At present, REEs are irreplaceable in technological development, they are used, among others, in computers, telephones, medical equipment, nuclear reactors and many other key solutions [40, 41]. Their role will also be indescribable in the energy transformation. Because rare earth elements occur in limited quantities and are concentrated in only a few countries on the globe, they have been classified as critical raw materials [42]. It has been determined that the currently known sources of REEs, assuming the intensive development of RES technology, will be sufficient in the world only until 2050, assuming that they will only be used in the EU and only in favour of wind energy [43]. In 2020, more than a quarter of the REEs obtained was used to produce permanent magnets, a large part of which was used in the production of electric vehicles and wind turbines [44]. Furthermore, most of the REEs available in global markets comes from a single supplier, i.e. China [45], which poses an additional threat to the success of the energy security and transition. In the case of energy security, a 30% share of a single source of raw material extraction [46] is considered acceptable for proper diversification, which is certainly not maintained in the case of REEs. Therefore, it is necessary to search for new sources and distribution channels for these elements. Rare earth elements are contained in hard coal in quantities insufficient for their extraction to be economically justified. However, the concentration of elements is much higher in fly ash, which is a waste product of the coal combustion process [47]. Taking into account also the elimination of costs associated with the exploitation of ores containing REEs, the possibility of obtaining many REEs from a single source, and the elimination of environmental costs and degradation of the natural environment, obtaining elements using effective ash recovery technology can be an excellent source of rare earth metals.

A serious problem for energy systems based on renewable energy sources is the lack of stable operation caused by the specificity of these energy sources. Their efficiency depends on weather conditions and the time of year and day. Obtaining energy only from these sources does not provide energy security, because in order to maintain it, energy must be supplied in the desired quantity, time and place, and at an acceptable price. The lack of flexibility of energy systems, e.g. in EU countries such as Poland, Hungary, Romania, means that when there is no possibility of using photovoltaics, these countries must rely on energy imports at times when it reaches the highest prices on the energy market [48]. To increase the share of renewable energy in the energy mix and at the same time achieve at least an acceptable level of energy security, it is necessary to support it with energy sources that can supply it regardless of weather conditions. Such a source are certainly fossil fuels, which, equipped with clean coal technologies, can provide clean energy when it is needed.

1.3 Challenges and Benefits of Using Clean Coal Technologies

Properly planned development and implementation of clean coal technologies could solve the problems of modern energy systems. To make this possible, it is necessary to develop appropriate documents and legal solutions. This applies, of course, not only to the European Union, but also to other countries that base their economy on coal and intend to rely on it in the future. The United States and China are currently among the countries that should be considered leaders in the development and implementation of CCT. Work on CCT is still ongoing, and China is making investments aimed at developing effective and efficient units for, among others, coal gasification or separation and use of greenhouse gases. China actively supports the development of CCT, using financial and legal instruments [49, 50], so that in the future coal, which is the basis of China's energy system, can continue to be used in an efficient and environmentally neutral way in the future. The United States mainly focusses on CCS technology, and thanks to financial support and legal regulations, it is the world leader in its development [51]. Full industrial implementation of CCT technology in the US also requires further work, financial resources, and the expansion of the necessary infrastructure.

The EU countries that should be primarily interested in clean coal technologies because of the size of their coal deposits are Poland and Germany. In recent years, interest in clean coal technologies has declined in the European Union due to the priority given to renewable energy sources and the goals set by the European Green Deal. The decline in the prices of renewable energy technologies also contributed to this. CCTs were also met with social resistance, which occurred, for example, in Germany [52]. However, in 2024, the European Union revised its coal policy. This was mainly due to the need to diversify energy sources in order to maintain the stable operation of the energy systems of the member states. Several factors contributed to this, mainly the war in Ukraine, problems with access to natural gas, slow development of renewable energy sources, and changes to the EU ETS system. As a result, the approach of the European Union has been slightly modified and the value of clean coal technologies, especially CCS and their role in ensuring access to clean energy generated from coal, has once again recognized. CCT should be a link between current energy systems and future systems based on renewable energy sources. In 2024, the European Commission presented a strategy for industrial management of carbon dioxide emissions, which includes support for the development of CCT and its promotion. However, the European Union will have to face the difficult challenge of making up for years of neglect and arrears on the subject of clean coal technologies. All countries interested in using CCT will have to intensify their actions in the field of their implementation on an industrial scale. The main challenges are developing a technically reliable technology, economically viable and which will also allow the management and use of waste generated during the use and combustion of fuel. However, the implementation of CCT will bring many benefits, first of all, it will enable clean combustion of coal, reduction of emissions not only of greenhouse gases, but also of other harmful substances generated during coal combustion. Some of them can also contribute to the optimisation of the combustion process and the improvement of energy efficiency.

References

1. Dodson J, Li X, Sun N, Atahan P, Zhou X, Liu H, Yang Z (2014) Use of coal in the Bronze age in China. The Holocene 24(5):525–530
2. Durucan S, Brenkley D (2010) Coal mining research in the United Kingdom: a historical review. In: Brune J (ed) Proceedings extracting the science—a century of mining research, pp 10–22
3. Oei PY, Brauers H, Herpich P (2020) Lessons from Germany's hard coal mining phase-out: policies and transition from 1950 to 2018. Clim Policy 20(8):963–979
4. Państwowy Instytut Geologiczny. Available online https://www.pgi.gov.pl/psg-1/psg-2/informacja-i-szkolenia/wiadomosci-surowcowe/10420-czy-wiecie-ze-wegiel-kamienny.html. Accessed on 15 Sept 2024
5. Martinez DM, Ebenhack BW (2008) Understanding the role of energy consumption in human development through the use of saturation phenomena. Energy Policy 36:1430–1435
6. BP Statistical Review of World Energy. Available online https://www.bp.com/en/global/corporate/energy-economics/statistical-review-of-world-energy.html. Accessed on 10 Sept 2024
7. The outlook for China's fossil fuel consumption under the energy transition and its geopolitical implications. Oxford Institute for Energy Studies (2023)
8. Eurostat (2024) Available online https://ec.europa.eu/eurostat/data/database. Accessed on 1 May 2023
9. RED: Renewable Energy Directive. Available online https://energy.ec.europa.eu/topics/renewable-energy/renewable-energy-directive-targets-and-rules/renewable-energy-directive_en. Accessed on 16 Sept 2024
10. Yan R, Ma M, Zhou N, Feng W, Xiang X, Mao C (2023) Towards COP27: decarbonization patterns of residential building in China and India. Appl Energy 352:122003
11. United Nations Framework Convention on Climate Change (2015) Paris agreement. Available online https://unfccc.int/sites/default/files/english_paris_agreement.pdf
12. UN (2024) The 2030 agenda for sustainable development. Available online https://sdgs.un.org/2030agenda. Accessed on 10 Sept 2024
13. Fetting C (2020) The European green deal. ESDN Report, 53
14. ScienceDirect (2024) Available online https://www.sciencedirect.com/. Accessed on 19 Sept 2024
15. BiznesAlert (n.d.) Available online https://biznesalert.pl/europa-wegiel-energetyka-usa-chiny-indie/. Accessed on 19 Sept 2024
16. Melikoglu M (2018) Clean coal technologies: a global to local review for Turkey. Energ Strat Rev 22:313–319

17. International Energy Agency (IEA) Clean Coal Centre (2016) Clean coal technologies. Available online http://www.iea-coal.org.uk/site/2010/database-section/clean-coal-technologies. Accessed on 11 Sept 2024
18. Chen W, Xu R (2010) Clean coal technology development in China. Energy Policy 38(5):2123–2130
19. OECD/IEA (1993) Clean coal technologies. Options for future
20. Vardar S, Demirel B, Onay TT (2022) Impacts of coal-fired power plants for energy generation on environment and future implications of energy policy for Turkey. Environ Sci Pollut Res 29(27):40302–40318
21. Blaschke W, Nycz R (2003) Clean coal-preparation barriers in Poland. Appl Energy 74(3–4):343–348
22. Baic I, Blaschke W, Gaj B (2019) Hard coal processing in Poland—current status and future trends. Sci Pap Inst Mineral Resour Energy PAS 108:83–97
23. Blaschke W (2008) Clean coal technologies begin with its enrichment. Energy Policy 11(2):7–13
24. Kadagala MR, Nikkam S, Tripathy SK (2021) A review on flotation of coal using mixed reagent systems. Miner Eng 173:107217
25. Nurhawayisah SR et al (2019) Screening of bacteria for coal benefit. In: IOP conference series: materials science and engineering. IOP Publishing, 012021
26. Hasan MMF et al (2015) A multi-scale framework for CO_2 capture, utilization, and sequestration: CCUS and CCU. Comput Chem Eng 81:2–21
27. Zhang L et al (2022) Frontiers of CO_2 capture and utilization (CCU) towards carbon neutrality. Adv Atmos Sci 39(8):1252–1270
28. Shi Y et al (2020) Energy and exergy analysis of oxy-fuel combustion based on circulating fluidized bed power plant firing coal, lignite, and biomass. Fuel 269:117424
29. Mandal R et al (2020) Laboratory investigation on underground coal gasification technique with real-time analysis. Fuel 275:117865
30. Mishra A, Gautam S, Sharma T (2018) Effect of operating parameters on coal gasification. Int J Coal Sci Technol 5:113–125
31. Asghar U et al (2021) Review on the progress in emission control technologies for the abatement of CO_2, SO_x and NO_x from fuel combustion. J Environ Chem Eng 9(5):106064
32. Rybak A, Rybak A, Joostberens J, Pielot J, Toś P (2024) Analysis of the impact of clean coal technologies on the share of coal in Poland's energy mix. Energies 17(6):1394
33. Chugh YP, Behum PT (2014) Coal waste management practices in the USA: an overview. Int J Coal Sci Technol 1:163–176
34. Dreger M, Kędzior S (2021) Methane emissions against the background of natural and mining conditions in the Budryk and Pniówek mines in the Upper Silesian Coal Basin (Poland). Environ Earth Sci 80:1–16
35. Wałachowska A, Ignasiak-Szulc A (2021) Comparison of renewable energy sources in 'New' EU member states in the context of national energy transformations. Energies 14(23):7963. https://doi.org/10.3390/en14237963
36. Chojnacka K, Wieczorek PP, Schroeder G, Michalak I (eds) (2018) Algae biomass: characteristics and applications: towards algae-based products, vol 8. Springer
37. Liu Y, Hou J, Zhao H, Liu X, Xia Z (2018) A method to recover natural gas hydrates with geothermal energy conveyed by CO_2. Energy 144:265–278
38. Zhong C et al (2023) Comparison of CO_2 and water as working fluids for an enhanced geothermal system in the Gonghe Basin, northwest China. Gondwana Res 122:199–214
39. Energy Dome (n.d.) Available online https://energydome.com. Accessed on 9 Sept 2024
40. Hardwerlibre. Hardwerlibre.com. Available online https://hardwerlibre.com. Accessed on 2 May 2024
41. Fernandez V (2017) Rare-earth elements market: a historical and financial perspective. Resour Policy 53:26–45
42. European Commission (2023) COM/2023/160. Available online https://eurlex.europa.eu/legalcontent/EN/TXT/?uri=CELEX%3A52023PC0160. Accessed on 2 May 2024

References

43. JRC. The role of rare earth elements in wind energy and electric mobility. Available online https://publications.jrc.ec.europa.eu/repository/bitstream/JRC122671/jrc122671_the_role_of_rare_earth_elements_in_wind_energy_and_electric_mobility_2.pdf. Accessed on 20 May 2024
44. Depraiter L, Goutte S (2023) The role and challenges of rare earths in the energy transition. Resour Policy 86:104137
45. Golev A, Scott M, Erskine PD, Ali SH, Ballantyne GR (2014) Rare earths supply chains: current status, constraints, and opportunities. Resour Policy 41:52–59
46. Rybak A (2020) Poland's energy security and coal position in the country's energy mix. Silesian University of Technology, Monograph 865, Gliwice
47. Rybak A, Rybak A (2021) Characteristics of some selected methods of rare earth elements recovery from coal fly ashes. Metals 11(1):142
48. WysokieNapiecie.pl. Upały w Europie, prąd w Polsce po 2000 zł. Available online https://wysokienapiecie.pl/102614-upaly-w-europie-prad-w-polsce-po-2000-zl/. Accessed on 3 Sept 2024
49. State Council of the People's Republic of China. Premier's News. Available online https://english.www.gov.cn/premier/news/202111/18/content_WS6195a201c6d0df57f98e51d8.html. Accessed on 6 Sept 2024
50. Ministry of Ecology and Environment of China. White Paper. Available online https://english.mee.gov.cn/Resources/publications/Whitep/202012/t20201222_814160.shtml. Accessed on 9 Sept 2024
51. U.S. Department of the Treasury. Inflation Reduction Act. Available online https://home.treasury.gov/policy-issues/inflation-reduction-act. Accessed on 12 Sept 2024
52. Clean Energy Wire. Quest for Climate Neutrality Puts CCS Back on the Table. Available online https://www.cleanenergywire.org/factsheets/quest-climate-neutrality-puts-ccs-back-table-germany#four. Accessed on 15 Sept 2024

Chapter 2
Coal Market

Abstract In this chapter, the coal market analysis was conducted in selected countries, such as China, the United States, and Poland. The availability of coal resources throughout the world and the structure of energy production systems selected for the analysis of countries were characterized. Machine learning algorithms, the Boltzmann, Gompretz, Holt-Winters models, and the random forest algorithm were used in creation of CFP 1.0 programme. It allowed to determine forecasts of the demand for energy from coal, CO_2 emissions, and coal prices on international markets. Factors from the SGD7 set that will have the greatest impact on the development of CCT were also identified. Alternative energy sources were indicated that could theoretically take over the role of coal in the near future.

Keywords Coal market · Machine learning · Energy demand

Coal is one of the most important energy carriers in the world. In 2022, its share in total global demand for primary energy was 27%, second only to crude oil (31%), which, however, has a different purpose and is mainly used in transport. Coal, in turn, is used to produce electricity and heat. The largest coal reserves in the world are located in the geographic area of the USA (29%), Russia (19%), China (16%), Australia (17%), and India (13%). In turn, the largest concentration of coal (3% of global reserves) in the European Union is located in Poland and Germany (Fig. 2.1).

Coal continues to play a key role in the energy mixes of countries, especially those with access to its resources. In China and India, coal provides 55% of primary energy consumption (Fig. 2.3), in Poland it is still around 42% (Fig. 2.2).

Coal consumption in China alone accounts for about 50% of the global demand for this fuel and is continuously growing, similarly to India. This is related to economic development, a large share of heavy industry in the industrial sector, population growth, and, at the same time, a rising standard of living of citizens. The share of coal in the energy mixes of countries around the world in the near future will be influenced by its impact on the natural environment. Many harmful gaseous substances are released during the combustion of fossil fuels, but 80% of them is CO_2, which is why it is of particular interest in terms of greenhouse gas emissions. In turn, 40%

© The Author(s), under exclusive license to Springer Nature Switzerland AG 2025
A. Rybak, *The Role of Clean Coal Technologies in Energy Transformation and Energy Security*, SpringerBriefs in Energy, https://doi.org/10.1007/978-3-031-80652-0_2

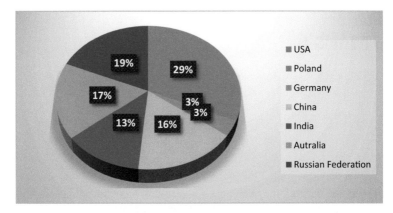

Fig. 2.1 World coal reserves in 2022, *source* own elaboration based on statistical data from [1]

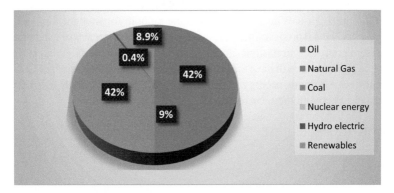

Fig. 2.2 Poland's energy mix in 2022, *source* own elaboration based on statistical data from [1]

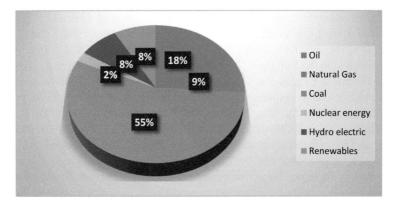

Fig. 2.3 China's energy mix in 2022 [6], *source* own elaboration based on statistical data from [1]

of the CO_2 emitted in the world is generated by the energy sector [2]. Therefore, countries around the world have undertaken the ambitious task of eliminating CO_2 emissions and, in accordance with the 2015 Paris Agreement, have developed their own documents regulating the issue of emission levels. The deadline to implement this undertaking varies depending on the country, so in the EU and the USA it is 2050, for China 2060 [3]. This results from the adopted plans and schedules that take into account the specificity of the energy systems and economies of individual countries. The European Union has also committed to reducing CO_2 emissions by 55% by 2030, which is included in the Fit for 55 strategy [4]. In turn, the USA wants to reduce its emissions by about 50% by 2030. It has also been assumed that the United States will decarbonise the energy sector by 2035 [5].

2.1 Market Forecasts for the Coal Sector

In order to verify the feasibility of implementing above-mentioned provisions, coal demand forecasts were made for China, the USA, and Poland by 2030. To analyse the coal market today, but also in a 5-year time horizon, machine learning algorithms were used. The Boltzmann and Gompretz sigmoid models and the Holt-Winters model were applied. The model parameters were optimised using machine learning algorithms. The random forest model was also used. CO_2 emissions and coal prices forecasts were also built.

The Holt-Winters model [6] was used to create CO_2 emission forecasts. This model can be applied to time series in which random fluctuations and trend were observed as components. The model parameters alpha and beta were optimised using an exhaustive algorithm that seeks the smallest value of the MAE model error [7].

The forecasts were calculated using the following formula:

$$F_t = \alpha \bullet Y_t + (1 - \alpha) \bullet (F_{t-1} + S_{t-1}) \tag{2.1}$$

where

F_t time series smoothed value,
S_t trend increment at moment t smoothed value,

Smoothing trend patterns over time t:

$$S_t = \beta \bullet (F_t + F_{t-1}) + (1 - \beta) \bullet S_{t-1} \tag{2.2}$$

$$\widehat{Y_t} = F_t + S_t m \tag{2.3}$$

where

$\widehat{Y_t}$ forecast of the dependent variable.

Sigmoid models, the SVM algorithm, and a random forest algorithm were used to determine the forecasts of the demand for energy from coal. Boltzmann and Gompretz models are functions taking the shape of the letter s describing the asymmetric growth and decline of time series values with constant inflection points [8]. The Boltzmann model allows to describe data whose values fall dynamically in the initial phase and then stabilize. The Gompretz model, on the contrary, describes time series where the growth for the initial observations is small, then increases dynamically and stabilises. These models are most often used in biology or medicine, but they are also perfectly suitable for describing economic data [9]. The models are described by the following equations:

Boltzmann [10]:

$$f_i(x) = \frac{(a_1 - a_2)}{1 + e^{\frac{(x_i - x_0)}{dx}}} + a_2 \qquad (2.4)$$

where

- a_1 horizontal asymptote of the function fi(x),
- a_2 horizontal asymptote of the function fi(x),
- x_i years,
- x_0 middle value of the interval,
- dx slope of the function.

Gompretz model [11]:

$$f_i(x) = A \times e^{-e^{-k(t-t_m)}} \qquad (2.5)$$

where

- A upper asymptote of the function $f_i(x)$,
- t time,
- k growth rate coefficient,
- t_m time at which the inflection of the function occurs.

Supervised machine learning was used. The programme written to determine the models requires input of initial values of the model parameters. They are then optimised by minimizing the square of the model residuals without having to programme them explicitly.

To build a coal demand model that would take into account many explanatory variables, a random forest algorithm was used [12]. It is a machine learning algorithm used to build a set of regression trees, thanks to which the results obtained are stable, resistant to overfitting and generalization of results [13]. The regression result is an average of the results obtained using individual trees. Model parameters are optimized based on minimizing the MSE error [14]. The model is described by the formula [15]:

2.1 Market Forecasts for the Coal Sector

$$\widehat{Y} = \frac{1}{q}\sum_{i=1}^{q}\widehat{Y}_i = \frac{1}{q}\sum_{i=1}^{q}\widehat{h}(X, S_n^{\Theta_i}) \tag{2.6}$$

where

$\widehat{Y}$ the result of the i-th tree,
i   1, 2, … q,
X   input vector with m features,
h   prediction function,
S_n training set containing n observations.

The normal distribution of the model residuals was confirmed by the Shapiro-Wilk test [16] and the Kolgomorov-Smirnov test [17]. Sigmoid models do not demand this kind of residuals testing, but the construction of confidence intervals requires that the residuals are characterized by a normal distribution. The lack of autocorrelation of the Holt-Winters model residuals was confirmed by the Durbin-Watson test. To verify the accuracy of the models, the Mean Absolute Percentage Error (MAPE) was used [18, 19].

Figure 2.4 presents the algorithm of the CFP 1.0 programme (Coal Future Predictor) written by the authors for the purposes of the research carried out. The WEKA [20] libraries were used for its creation. The programme initiates the algorithm selected by the user and collects the entered data. In the next step, the initial selected model parameters are optimised. A model with optimal parameters is built and the model residuals are verified. If case they are positively verified in terms of normality of distribution, the MAPE error and the forecast confidence interval are determined. The programme prints the performed calculations results in the console and ends the work.

Figures 2.5, 2.6 and 2.7 present the actual values of coal energy demand in the years 2000–2022 for China, the USA, and Poland. The forecast was made using the Gompretz and Boltzmann models. Two scenarios were also created for the forecast. It should be noted that coal demand in China is systematically growing. This is the result of intensive economic development, heavy industry, population growth, and ongoing urbanisation. Turbulence on the natural gas market, problems with access to raw material and volatile natural gas prices have prompted China to return to coal and increase its share in the energy system. The pace of renewable energy development was also not sufficient to meet China's growing energy needs. In connection with China's climate commitments by 2050, it was assumed that coal demand should stabilise at a constant level in the coming years. Therefore, the Gompretz model was used, which guarantees that the variable will tend to the upper asymptote of the model. By 2030, the demand according to the forecast (scenario 1) will increase by about 2%. Two scenarios were also created using the confidence interval. The confidence interval is the forecast range in which the value will lie with 95% probability. In Scenario 2, it was assumed that the demand for coal will be approximately 10% higher than the forecast, which will be influenced by, for example, limited development of RES, constantly growing demand for energy, and use of CCT. In Scenario 3, the

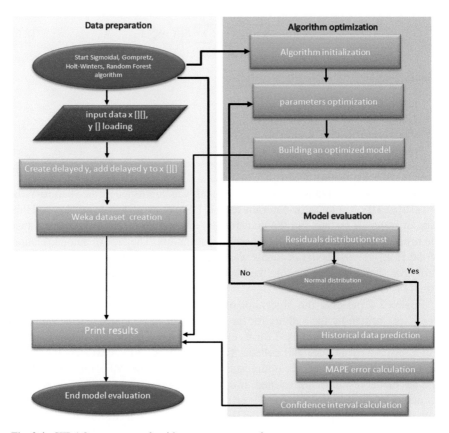

Fig. 2.4 CFP 1.0 programme algorithm, *source* own study

demand for coal will be 10% lower, for example, due to intensive development of RES.

In case of USA and Poland, the Boltzmann model was used because in both cases the time series are characterised by a downward trend. The analysis was again carried out for the years 2000–2022 with a forecast until 2030. In this time range, in the case of the USA, a complete course of the s-shaped model can be observed, because this period an intensive decline in the share of coal in the energy mix occurred. In the years 2000–2022, demand dropped by 115%.

In the case of Poland, the turning point, when coal consumption fell rapidly, occurred in the 1990s, and therefore the decline is not that clear after 2000. In the period under review, demand fell by 30%.

In both cases, this change was initiated by the climate policy of the European Union and the United States, which aims to decarbonise the energy sector.

Also, for the USA and Poland, two scenarios were made, the first of increased demand for energy from coal and the second, where the demand is lower than the forecast. For the USA, these scenarios differ by 35% from the forecast, in the case of

2.1 Market Forecasts for the Coal Sector

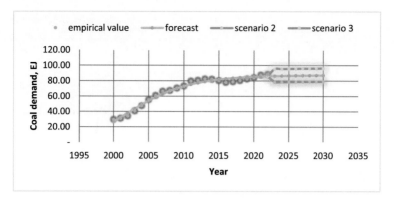

Fig. 2.5 Coal demand in China empirical values, forecast, scenarios until 2030, *source* own elaboration

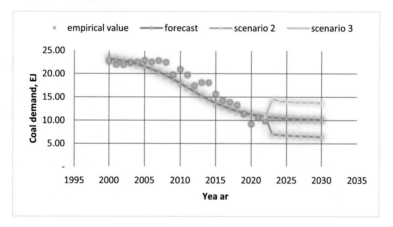

Fig. 2.6 Coal demand in the USA, empirical values, forecast, scenarios until 2030, *source* own elaboration

Poland by about 20%. The decrease in demand for coal energy may be caused by the increase in the share of RES in energy mixes, while the increase may be caused by the lack of possibility to ensure stable operation of the energy system through RES, a too slow pace of their development, and the use of CCT.

The main global factor that led to the decision to withdraw coal from energy mixes is the emission of greenhouse gases, which occurs during fuel combustion. Combating CO_2 emissions in the European Union is one of the most important goals achieved through the implementation of climate strategies, mainly EGD. The EU ETS CO_2 emissions trading system was also built for this purpose. The EU aims to become a leader in achieving energy neutrality goals. In connection with this, it imposes ambitious goals on its member states, such as reducing emissions by 55% by 2030. This is a particularly complicated task for countries that are still highly

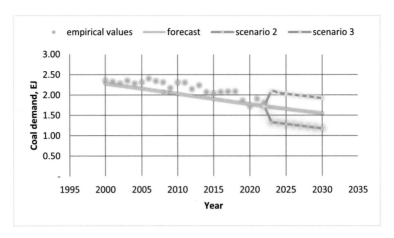

Fig. 2.7 Coal demand in Poland, empirical values, forecast, scenarios until 2030, *source* own elaboration

dependent on coal. Poland is the best example here. An intensive decrease in CO_2 emissions occurred in Poland until 1994, which resulted from the change of the political system and the transition to a market economy. From 1989 to 1994, CO_2 emissions fell by 34%. However, this was caused by the reduction and liquidation of energy-intensive industries. After this period, stagnation occurred and annual CO_2 emissions oscillate around 300 million Mg. Accelerating CO_2 reduction currently requires the use of new organisational, legal and technological solutions.

The Holt-Winters model allowed the construction of a time series model of CO_2 emissions in Poland, the USA, and China in the years 2003–2022 together with a forecast until 2025. The MAPE errors of the models did not exceed 4%. According to the forecast built for Poland, by 2025, a 1% decrease in emissions can be expected compared to the last known observation (Fig. 2.8).

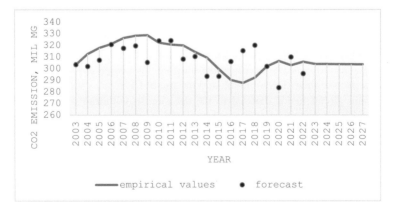

Fig. 2.8 CO_2 emissions in Poland in 2003–2022 with forecast, *source* own elaboration

2.1 Market Forecasts for the Coal Sector

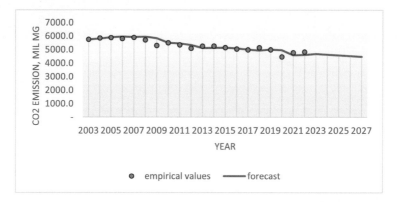

Fig. 2.9 CO_2 emissions in the USA from 2003 to 2022 with forecast, *source* own elaboration

In the USA, CO_2 emissions began to decline in the years 2009–2015 and this decrease amounted to approximately 10%. The main reason in this case was the shale revolution, where the obtained fuel made it possible to replace coal in the energy production structure [21]. This was also the result of the increase in energy efficiency and the development of renewable energy sources. The financial crisis that took place in the USA in 2008 also contributed to the reduction of energy demand and thus to the decrease in CO_2 emissions [22]. The reason was also the use of modern technologies in heating, industry, and energy. After 2016, emissions began to increase slightly as a result of economic growth, as well as changes in climate policy introduced by President Trump [23]. The forecast made indicates that by 2025 emissions will fall by 3% compared to the last known observation (Fig. 2.9).

In China, CO_2 emissions are continuously increasing, mainly caused by intensive economic growth, increasing demand for energy in industry for electricity. This causes China to continue to rely on coal and to build new coal-fired power plants [24]. The forecast shows that CO_2 emissions in China will continue to increase and may increase by another 2% by 2025 (Fig. 2.10).

In 2023, coal prices were rising due to the war in Ukraine. Sanctions imposed on the Russian Federation resulted in a reduction in the possibility of importing fuel from that direction. Additionally, limited gas supplies resulted in a return to coal as a substitute fuel. Rising natural gas prices also strengthened interest in coal. At the same time, unfavourable weather conditions, especially in Australia, disrupted supply chains in that direction, which caused price increases (Fig. 2.11).

Coal prices described by the Polish PSCMI index were shaped similarly (Fig. 2.12). At the beginning of 2024, coal prices fell on world markets mainly due to reduced demand on Asian markets, where large amounts of water energy were obtained as a result of heavy rains [25]. The renewed increase in coal prices was mainly influenced by demand on Asian markets, as well as the war in the Middle East. This shaped the value of the CIF ARA, FOB RB, and PSCMI index.

Coal prices are generally more favourable than natural gas and oil, but its attractiveness in the future will be mainly influenced by CO_2 emissions and enablers for

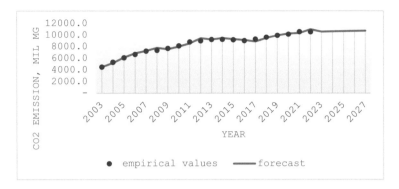

Fig. 2.10 CO_2 emissions in China from 2003 to 2022 with forecast, *source* own elaboration

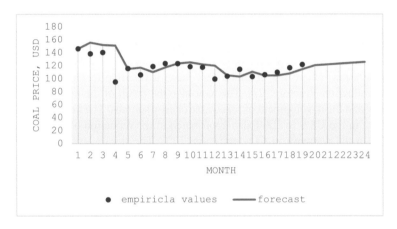

Fig. 2.11 Coal price (API2) CIF ARA (ARGUS-McCloskey) futures (MTFc1), 1/2023–7/2024, *source* own elaboration

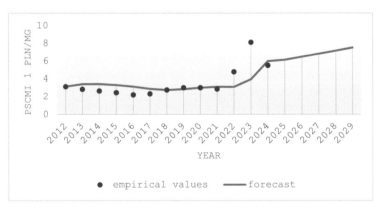

Fig. 2.12 Polish PSCMI index in 2012–2024 with forecast, *source* own elaboration

2.1 Market Forecasts for the Coal Sector

achieving decarbonisation and sustainable development goals. Therefore a model of demand for energy from coal in Poland was built, taking into account 9 explanatory variables (Fig. 2.13). Independent variables were the SDG 7 target implementation measures determined by Eurostat. The importance of the influence of individual explanatory variables on the demand for energy generated by coal was determined. A multiple regression model was used for this purpose. The greatest influence is exerted by energy productivity, the demand for energy, CO_2 emissions, the share of renewable energy sources, and dependence on energy imports (Table 2.1).

A random forest machine learning algorithm was used. The MAPE error of the model was 1.2%. The calculated forecast indicates that by 2030 demand for coal will be 1% higher than in 2022. The confidence interval specifies the distribution of uncertainty of the forecasts made. It was constructed for a confidence level, between the 2.5 and 97.5% percentiles. The confidence interval indicates that the use of coal

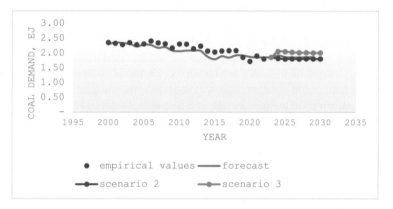

Fig. 2.13 Demand for coal energy in Poland, actual values, forecast to 2030 and demand scenarios, *source* own elaboration

Table 2.1 The importance of factors influencing coal demand

Indicator	Weight
Primary energy consumption, Mtoe	0.22
Final energy consumption, Mtoe	0.04
Final energy consumption in households per capita, KGOE	0.00
Energy productivity, KGOE	0.29
Share of renewable energy in gross final energy consumption by sector, %	0.15
Energy import dependency, %	0.11
Energy import dependency—solid fossil fuels, %	0.03
Population unable to keep home adequately warm by poverty status, %	0.00
CO_2 emission, Mg	0.16

may be 10% higher than predicted and 2% lower than indicated by the forecast (Fig. 2.13).

In conclusion, the forecasts performed indicate that by 2030, in the case of the United States and Poland, no large decreases in demand for energy generated from coal should be expected. The forecast performed indicates a stabilisation of demand, therefore, if the analysed phenomenon continues to be affected by the factors that have influenced the time series so far, coal consumption in Poland will be at the level of about 1.5 EJ ± 0.5 EJ, in the USA 10 EJ ± 2 EJ. In order to change this, it will be necessary to take additional decisive actions to quickly eliminate coal. The forecast using the random forest model, taking into account SDG7 measures as explanatory variables for Poland, gave similar results and even indicates a greater demand for coal. This is also visible in the case of scenarios. Scenario 3 assumes an increase in demand for energy from coal by 11%, and scenario 2 a decrease in demand by about 3%. In the case of China, on the other hand, demand is still growing. This is consistent with the results of the forecasts on CO_2 emission levels. In China, continuous increases in emissions should be expected until 2030; in the USA and Poland it will decrease, but at too slow pace. Coal prices are generally more favourable than other fossil fuels. In connection with the above, it was observed that coal will most likely remain in the energy mixes of the analysed countries. Therefore, it is necessary to apply technical measures and solutions that will accelerate the elimination of greenhouse gas emissions, which will also affect the decrease in fuel prices due to the elimination of emission costs.

2.2 Alternative Energy Sources

Given the complicated political situation in the EU and the military threat that has been present for two years, the assumptions regarding the way to decarbonise the energy sectors of the member states should be reconsidered. In addition to coal, EU countries also use other fossil fuels with similar environmental impacts, especially oil, but also natural gas. The transformation process of energy systems is to be based on renewable energy sources.

China, the USA, Poland, and other countries that still use coal in the majority are taking care of the development of renewable energy sources, but these carriers still face huge challenges, especially in terms of sufficient infrastructure development and storage of acquired solar and wind energy. In order to achieve neutrality, China, the USA and Poland plan to use mainly wind and solar energy, as well as biomass. There are also plans to develop nuclear energy. Energy storage is also an important issue. The key challenge will be the industrial and transport sectors, which are strongly associated with fossil fuels. The use of clean coal technologies will be important in this respect. It is necessary to carry out appropriate legislative changes in energy policy, which are to encourage action in accordance with the guidelines of the energy transformation. However, this is an extremely complicated task, therefore, there are concerns that these countries will not be able to achieve the set goals. In the case of

China, it will also have an impact on global goals, taking into account its share in the global demand for coal and greenhouse gas emissions [26].

References

1. BP Statistical Review of World Energy. Available online https://www.bp.com/en/global/corporate/energy-economics/statistical-review-of-world-energy.html. Accessed on 10 Sept 2024
2. IEA—International Energy Agency. Available online https://www.iea.org/. Accessed on 2 Sept 2024
3. Lu Q, Duan H, Shi H, Peng B, Liu Y, Wu T, Wang S (2022) Decarbonization scenarios and carbon reduction potential for China's road transportation by 2060. NPJ Urban Sustain 2(1):34
4. European Commission. Available online https://eur-lex.europa.eu/legal-content/PL/TXT/?uri=CELEX%3A52021DC0550. Accessed on 5 Sept 2024
5. Vardinov P, Milne J (2022) USA 2030: national energy strategy. 7159EXQ: energy economics and policy
6. Chatfield C, Yar M (1988) Holt-winters forecasting: some practical issues. J R Stat Soc Ser D (Stat) 37(2):129–140
7. Chai T, Draxler RR (2014) Root mean square error (RMSE) or mean absolute error (MAE). Geosci Model Dev Discuss 7(1):1525–1534
8. Gompertz B (1825) On the nature of the function expressive of the law of human mortality, and on a new mode of determining the value of life contingencies. Philosoph Trans R Soc Lond B Biol Sci 182:513–585
9. Tjørve KMC, Tjørve E (2017) The use of Gompertz models in growth analyses, and new Gompertz-model approach: an addition to the unified-Richards family. PLoS ONE 12(6):e0178691
10. Milanov P et al (2018) Curve fitting problem: torque–velocity relationship with polynomials and Boltzmann sigmoid functions. Acta Bioeng Biomech 20(1):169–184
11. Cao L, Shi P-J, Li L, Chen G (2019) A new flexible sigmoidal growth model. Symmetry 11(2):204
12. Wang F, Wang Y, Zhang K, Hu M, Weng Q, Zhang H (2021) Spatial heterogeneity modeling of water quality based on random forest regression and model interpretation. Environ Res 202:111660
13. Breiman L (2001) Random forests. Mach Learn 45:5–32
14. Vivas E, Allende-Cid H, Salas R (2020) A systematic review of statistical and machine learning methods for electrical power forecasting with reported map score. Entropy 22(12):1412
15. Li Y, Zou C, Berecibar M, Nanini-Maury E, Chan JCW, Van den Bossche P, Omar N (2018) Random forest regression for online capacity estimation of lithium-ion batteries. Appl Energy 232:197–210
16. Shapiro SS, Wilk R (1965) An analysis of variance test for normality (complete samples). Biometrika 52(3–4):591–611
17. Yazici B, Yolacan S (2007) A comparison of various tests of normality. J Stat Comput Simul 77(2):175–183
18. Bliemel F (1973) Theil's forecast accuracy coefficient: a clarification. J Mark Res 10(4):444–446
19. Farnum NR, Stanton W (1989) Quantitative forecasting methods. PWS-Kent Publishing Company, Boston
20. Waikato. Weka software. Available online https://waikato.github.io/weka-wiki/downloading_weka/. Accessed on 8 Sept 2024
21. Salygin V, Guliev I, Chernysheva N, Sokolova E, Toropova N, Egorova L (2019) Global shale revolution: successes, challenges, and prospects. Sustainability 11(6):1627

22. Swedberg R (2013) The financial crisis in the US 2008–2009: losing and restoring confidence. Soc Econ Rev 11(3):501–523
23. Collins SD (2020) America first: the trump effect on climate change policy. Non-human nature in world politics: theory and practice, 179–203
24. Weng Z, Song Y, Cheng C, Tong D, Xu M, Wang M, Xie Y (2023) Possible underestimation of the coal-fired power plants to air pollution in China. Resour Conserv Recycl 198:107208
25. Polski Rynek Węgla. Międzynarodowy rynek węgla w czerwcu 2024 r. Available online https://polskirynekwegla.pl/miedzynarodowy-rynek-wegla-w-czerwcu-2024-r. Accessed on 10 Sept 2024
26. Luo S, Hu W, Liu W, Zhang Z, Bai C, Huang Q, Chen Z (2022) Study on the decarbonization in China's power sector under the background of carbon neutrality by 2060. Renew Sustain Energy Rev 166:112618

Chapter 3
Clean Coal Technologies (CCT)

Abstract This chapter characterizes the most important and promising clean coal technologies such as coal gasification in reactors and in the bed, oxy-combustion, supercritical power plants, CCS, CCU, CCUS, and innovative membrane techniques. It also describes the possibilities and methods of waste management generated in the coal combustion process. The waste is divided into two categories, gaseous waste such as SO_x, NO_x, CO_2, and solid waste, mainly fly ash. The possibilities of waste management were also presented.

Keywords Clean coal technologies · CO_2 sequestration · Membrane techniques · Circular economy · Rare earth elements

The most promising clean coal technologies used in the coal combustion stage include, among others, coal gasification, oxy-combustion, supercritical power plants with steam boilers and fluidized bed boilers [1]. Methods are also used that enable the use of traditional combustion methods but provide the possibility of cleaning the generated flue gases. They can be implemented before or during the combustion process.

Combustion in an oxygen-enriched atmosphere involves burning fuel in an atmosphere of pure oxygen. This improves the efficiency of the fuel combustion process. The combustion process can also be carried out with recirculation of exhaust gases. As a result of this process, a gas mixture is created during combustion, where almost 100% is carbon dioxide ready for the sequestration process [2]. Because the combustion process takes place without access to nitrogen, no nitrogen oxides are produced. Despite its advantages, this process can pose an explosion hazard due to the contact of oxygen with the powdered fuel [3]. Work on oxy-combustion technology has been carried out in many countries, including Germany in the Schwarze Pumpe project, in Australia as part of the Callide project, or in Spain in the CIUDEN project [4, 5].

Another solution is to burn coal under supercritical conditions. This involves generating a so-called supercritical phase in the boiler by heating water to a temperature above the critical value, i.e. a temperature of 374.15 °C and pressure of 22.11 MPa. This creates a coherent phase of water and steam at the same temperature

[6], which allows efficient energy generation and limits the amount of CO_2 emitted. The disadvantage of this solution is certainly the costs associated with the construction and maintenance of the installation, which must be constructed of appropriate materials resistant to high temperatures, pressure, and corrosion. Currently, supercritical technology is used in power plants around the world, in the USA, China, Poland, Japan, and Germany. Modifications of this technology are also used, such as Ultra-Supercritical USC technology, where temperatures and pressure reach even higher values, resulting in greater efficiency of the combustion process [7].

Coal gasification involves the conversion of fuel into syngas during a chemical reaction carried out in a reactor or furnace. It consists mainly of carbon monoxide, hydrogen, carbon dioxide and nitrogen [8]. Gasification can be carried out in various ways, e.g. using air, carbon dioxide, oxygen and steam, or a set of reagents [9, 10]. Additionally, depending on the place where the process occurs, it can be divided into underground (in a bed) and above-ground gasification. Different reactors can be distinguished, such as fluidized bed, and dispersion, moving bed, or transporting reactors. Reactors can also be divided according to the generation to which they are classified. Generation I includes reactors that are rarely used and are falling into disuse, which were introduced at the beginning of the twentieth century. These are moving-bed (Lurgi), dispersion (Koppers-Totzek), or fluidized bed (Winkler) reactors. Generation II includes the reactors that are the most commonly used, and the technology can be considered mature. These include, for example, HTW ConocoPhilips, Shell, Texaco, or Siemens dispersive reactors [11]. Generation III, on the other hand, includes reactors whose technology has been developed since the end of the twentieth century, such as KBR transport reactors or Mitsubishi dispersive reactors [12]. Generation III reactors will play an important role in the process of decarbonisation and shaping sustainable energy systems.

In a fluidized bed reactor, coal particles float in a mixture of carrier material, e.g. sand and ash. The bed particles are suspended in a stream of supplied gas–oxygen, steam, or air. The reactor can be circulating or stationary. Fluidized bed reactors can be used for various fuels. The combustion process is uniform and therefore effective [13]. They are becoming increasingly popular, but dispersion reactors are the most popular. A dispersion reactor is one in which the reactant is dispersed into small particles in another gas or liquid. This increases the surface area on which the gasification reaction can take place, conducted at a very high temperature. The process is characterised by high efficiency and purity of the resulting gas [14]. A moving-bed reactor is a specific type of reactor that is fed with fuel from above. The fuel moves downward by gravity, while oxygen, CO_2 or steam are supplied in the opposite or synchronous direction to the movement of the bed. Due to this, the process is effective and its precise control is possible [15]. Transport reactors are a form that combines the advantages of a fluidized bed and a dispersion reactor. Carbon particles float in the process gas stream, causing a combustion reaction to occur and the temperature to rise. The gas and cabronizate produced are transported to the upper part of the reactor, where gasification takes place [16].

These reactors can also be used in advanced coal gasification systems such as Integrated Gasification Combined Cycle IGCC. After the gasification process in the reactor, the syngas produced is purified and used to generate electricity [17]. This allows for high process efficiency, effective reduction of pollutant emissions, and integration of IGCC with CCS. IGCC technology is considered a solution that will replace conventional coal-fired power plants and enable the energy transition process.

In addition to above-ground gasification, it is also possible to carry out this process directly in the coal deposit [18]. This is particularly important in the case of deposits whose exploitation is impossible due to geological conditions or unprofitable. Underground Coal Gasification UCG is a solution that was started at the beginning of the twentieth century [19]. The concept started by W. Williams in 1868 and continued by D. Mendeleev has found interest all over the world [20]. The first attempt at underground gasification was undertaken in 1912 in Durham [21]. Then, research interrupted by World War I was continued in the USSR, where the first commercial UCG installation was built in 1937. In the 1970s, a UCG development programme was created in the United States, but at the end of the twentieth century, interest in UCG has faded. In the twenty-first century, interest in this topic has returned, especially in the context of the growing demand for energy obtained without emitting greenhouse gases. Numerous development projects have been carried out in Australia, China, Poland, and South Africa [22–24]. The technology is based on drilling wells leading to the coal seam. An oxidiser is supplied to the deposit through the boreholes. The deposit is ignited, and syngas is produced, which is extracted to the surface and purified by a production well. Due to the numerous problems associated with the conductivity of the underground gasification process and its control, numerous projects have been abandoned. There are many underground gasification methods, shaft UCG methods such as chamber or warehouse method, Borehole producer method, stream method, LLT gasification method and shaftless UCG methods such as one of the oldest LVW and controlled retractable injection point [25–27].

The syngas produced during gasification is subjected to further processing. One of the possibilities of its use is its transformation into liquid hydrocarbons. Due to the access to crude oil, this process is not used on a global scale, but it was exploited in times of limited access to liquid fuels, e.g. during World War II or in South Africa during the oil embargo in the 1970s [28]. Fischer–Tropsch (FTS) synthesis is a technology that, because of a series of chemical reactions taking place under precisely controlled temperature and pressure conditions and in the presence of catalysts (e.g. based on iron, cobalt), allows the transformation of the synthesis gas into liquid hydrocarbons. The FTS technology is mature and available from corporations such as Shell, Sasol-Chevron, and BP [29].

To avoid the emission of mainly greenhouse gases during coal gasification, the process can be enhanced with a carbon capture and storage (CCS) installation. CCS can also be used in traditional coal-fired power plants.

The CO_2 sequestration process provides an opportunity to transform current energy systems into low-emission and decarbonisation-supporting ones. There are 186 companies around the world that deal with CO_2 capture, and interest in projects related to this topic is constantly growing, including in the EU. This is largely due to the recently increased prices of EU ETS allowances, which have increased by as much as 5 times in 2021–2023 [30]. Poland is also taking action to develop CCS, and its potential for storing CO_2 produced in the energy sector would be enough for 100 years [31]. Most CCS projects around the world are implemented in the United States, Canada, China, Australia, and Norway [32, 33]. The CCS installation can be integrated with the existing energy system, which is a major advantage of this solution. Sequestration, depending on the place of storage of the collected CO_2, is divided into geological, where CO_2 is stored in rock formations, for example after depleted oil or natural gas deposits, storage in the seas, where CO_2 is introduced to great depths, or mineral sequestration, when CO_2 is subjected to a reaction with minerals rich in carbonates. The CO_2 capture process can take place before, during, or after combustion of the fuel. Subsequently the captured CO_2 is usually transported to the storage site by pipeline. The disadvantage of CCS is the impact of its use on the efficiency of the energy production process. CCS requires access to energy, and investments related to CCS increase the costs of building an energy installation by approximately 30%, which affects the cost of energy production [34].

Modification of CCS technology is Carbon Capture and Utilisation (CCU) and CCUS Carbon Capture, Utilization and Storage. These technologies are designed to enable CO_2 capture and its subsequent use, for example, in the process of extracting crude oil (EOR), synthesis of fuels, other chemicals, or construction materials. CO_2 is separated from gases produced by the industry using various technologies, e.g., absorption or membrane techniques. Many CCUS projects have been carried out in the world, e.g. in Canada or Norway [35].

Membrane techniques should be considered a promising method of carbon dioxide capture. Various types of membrane have been developed that can be used in the process of separating greenhouse gases. One of them are hybrid membranes. The membrane must be characterised by appropriate physical and chemical parameters so that the process can proceed in an effective and undisturbed manner. Hybrid membranes allow for the appropriate selection of components, due to which membrane techniques are characterised by continuity of operation, low energy consumption, and ease of modification and adaptation to a specific industrial process.

3.1 Membrane Techniques

CO_2 separation can be carried out using various conventional methods, such as adsorption, chemical combustion, absorption, cryogenic distillation, and hydrate-based separation [36]. However, most of proposed conventional methods are very energy-consuming and expensive.

A promising alternative to conventional techniques can be membrane methods, which usually have many advantages, such as high performance, low energy consumption, no need to add chemicals, i.e. no waste streams, simple constructions, ease of up-scaling, conducting separation in a continuous manner, possibility to easily combine membrane processes with other unit processes (hybrid processes),

3.1 Membrane Techniques

the possibility of improving the separation properties of membranes during the operation of the system, conducting separation in mild environmental conditions, etc. In the case of separation of gas mixtures, several crucial factors should be considered, such as improved gas permeability and selectivity, high mechanical resistance as well as thermal and chemical stability [37].

In the separation of gas mixtures, among others, various types of membranes, both organic and inorganic, are used for separation of CO_2. Each of these groups of membranes has both disadvantages and advantages. For example, polymer membranes are characterized by high permeability, they can be obtained in the form of thin membranes but have low thermal and chemical stability. While inorganic membranes are usually characterized by high mechanical, thermal and chemical strength. However, they require significant financial outlays and are characterized by lower processing capacity than organic polymer membranes [38]. So, the main goal of the current research is to create a highly permeable and selective membranes with appropriate strength. Therefore, the introduction of new polymer materials for gas separation or modification of existing ones has become a particularly crucial factor in the development of membrane techniques.

In the case of separation of a gaseous mixture, using polymer membranes you can only get some limited selectivity and permeability, that is why polymers should be modified by various substitutions, such as: carboxylation, bromination, acylation, sulfonylation, sulfonation, etc. All these processes can improve their chemical and thermal strength and selectivity of the examined membranes, but unfortunately can lead to a reduction in permeability for separated gas components [39]. Such substituted polymers can be modified further by introducing metal cations, such as: Na^+, Mg^{2+}, Al^{3+}. It is also possible to create copolymers or carbon membranes obtained by pyrolysis of polymer precursors (phenolic resins, polyimides, polyfurfuryl alcohol, cellulose, polyacrylonitrile, polyphenylene oxides, etc.) [40]. In order to overcome the limitations of polymer membranes and inorganic membranes, research is still being conducted to create alternative membrane materials.

Therefore, the next, very promising strategy that allows you to improve the separation properties and transport of gases through polymer membranes is the introduction of inorganic additives (zeolites, carbon molecular sieves, silica nanoparticles, carbon nanotubes, silicates, metal nanoparticles, metal oxides, carbon black, carbon fibers, metal organic framework, graphene, etc.) to the polymer matrix. This new class of membrane materials with significant potential in membrane separation technology is called mixed matrix membranes (MMMS) or hybrid membranes. In many cases, the polymer matrix plays the role of a dispersion medium here, which stabilizes and protects micro- and nanoparticles, while fillers introduced to the polymer positively modify their mechanical, optical, or electric properties. These membranes combine the advantages of individual phases, such as the selectivity of the filler particles and the ease of obtaining polymer membranes [41]. They can be divided into three main groups, such as: the polymer-solid, the polymer-liquid and the polymer-liquid–solid membranes. However, the most interest aroused the polymer-solid membranes. As a polymer matrix for the production of hybrid membranes, both glassy and rubber polymers were used. In the case of rubber polymers, the interfacial interaction between them and fillers is probably stronger due to the greater mobility of polymer chains [42]. On the other hand, glassy polymers were characterized by greater mechanical stability and better gas transport properties, but unfortunately, they often showed the presence of interfacial void defects [43]. Reduction of potential problems related to the defects of interfacial surfaces can be achieved by many methods, such as application of silane and amine coupling agents, coating of nonselective slots with a silicone rubber, addition of plasticizers etc. [44].

Thus, it can be seen that the effectiveness of this type of membranes depends mainly on the choice of polymer matrix, inorganic filler, and also on the interaction between these two materials, their compatibility and good dispersion of the inorganic phase [45].

It was found that by dispersing impermeable filler particles in polymers can reduce the gas permeability, which may be caused by an increase in the diffusion pathway tortuosity across the polymer film and to a loss of penetrant gas solubility [42].

However, many tests also showed the existence of a reverse trend in the case of glassy polymers, filled with non-porous fillers, which have predispositions to form aggregates, which leads to a change in the packing of polymer and creating void spaces [46]. These types of functional inorganic–organic hybrid composites have become materials of great interest of researchers, because their properties can be properly adjusted by controlling: composition, content and morphology of added inorganic filler particles, usage of different processing techniques or modification of polymer matrix (improvement of interaction between these two materials, their compatibility and dispersion of the inorganic phase) [47].

3.2 Waste Management from the Coal Production and Combustion Process and Its Role in the Development of Renewable Energy Sources

Waste is generated along the entire coal fuel processing chain, from the moment of exploitation, through the combustion process and energy generation. It has various forms, solid, such as coal waste accompanying coal seams, waste from the coal enrichment process, waste from the coal combustion process in solid form, i.e. slag, mill waste, and fly ash. Solid waste can be used in construction, backfilling of excavations, and land reclamation [48]. Furthermore, the coal combustion process generates gaseous waste, mainly CO_2, as well as SO_X, and NO_X [49].

3.2.1 Solid Waste

The largest part of the solid waste is fly ash. It consists mainly of Al_2O_3, CaO, SiO_2, Fe_2O_3, and Na_2O_2. Ash is generated in huge amounts, for example 4 million Mg/year in Poland [50] and over 500,000,000 Mg worldwide, of which only 30% is utilized [51, 52]. The remaining part must be collected in landfills. This is associated with storage costs and may additionally have a negative impact on the natural environment. Ash is subject to dusting and penetrating the soil may lead to its contamination and groundwater contamination through heavy metals such as As, Cd, Hg, Ni, Pb, Cr, Sr, Be, V, and U [53]. They also include rare earth elements (REEs). REEs are contained in coal but in quantities too small to be cost-effective to extract. However, after burning the organic substance, a concentrate of REEs contained in the ash is created. REEs is a group of 17 elements of strategic and critical importance [54]. They are one of the pillars of the modern world, used in electronic, industrial, and energy technologies. Without REEs, computers, mobile phones, wind turbines, electric motors, LCD screens, etc. would not exist [55]. Furthermore, rich REEs deposits are very irregularly distributed throughout the world, which is why they are mainly concentrated in the geographical area of China, the United States, Brazil, Vietnam, Australia, India, and the Russian Federation [56]. Deposits in the European Union are small and REEs production is negligible. China has 36% of the world's resources, but it is the largest producer and accounts for about 70% of production. They supply 98% of REEs consumed in the EU [57]. The growing demand for REEs has an impact on the availability and price fluctuations of the elements. In the years 2009–2011, the price of REEs increased fivefold, while currently we are dealing with price drops caused by overproduction in China. In the face of the growing demand for REEs, China is using a strategy aimed at discouraging competition, mainly the USA, which in turn is trying to regain the leading position in the REEs market lost at the end of the twentieth century. Such market tactics can put REEs recipients in a situation of REEs shortages overnight, which will be crucial to the success of the energy transition of their energy systems. Therefore, it is necessary to search for new sources of REEs.

Fly ash from Polish coal contains mainly light elements such as neodymium, cerium and praseodymium in the amount of 300 ppm and heavy elements such as terbium, dysprosium, erbium, yttrium, ytterbium, lutetium, holmium and thulium in the amount of 150 ppm [58, 59]. The authors calculated that REEs obtained from Polish ash produced in one year could cover at least 100% of the annual demand for wind energy.

Figure 3.1 presents the value and amount of REEs recovered from fly ash generated in Poland during one year. The largest amounts that can be recovered from fly ash are neodymium 248 Mg/year, and the smallest amounts are praseodymium 90 Mg/year. The highest value is achieved by the recovery of Dy and amounts to 90 million USD/year. In total, the recovered elements reach a value of about 160 million Mg/year.

Figure 3.2 presents the determined demand coverage index using recovered REEs. It is the lowest in the case of DDPMG technology for Tb and amounts to 1.2 (120%), and the highest for PMG1G and praseodymium technologies, where it reaches a value of 25.7.

The resulting surplus production can be made available to other EU countries, and the gained financial resources can be used for further development and work on CCT.

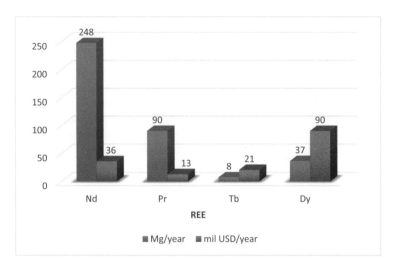

Fig. 3.1 Value and amount of REEs recovered from fly ash generated in Poland during one year, *source* own elaboration

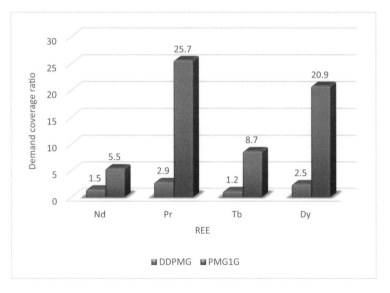

Fig. 3.2 Demand coverage ratio of REEs recovered from fly ash generated in Poland during one year, *source* own elaboration

3.2.2 Gaseous Waste

The gaseous waste generated during the combustion of fossil fuels, including coal, is primarily CO_2, NO_x, and SO_x. These gases can be separated from the exhaust gases, e.g., using CCS technology for carbon dioxide. However, transport to the storage site and the storage of gas itself are cumbersome and expensive. Therefore, a better solution is CCU, i.e. gas separation and its use. Section 1.2 described the possibility of using CO_2 to store energy generated by RES. CO_2 can also support oil extraction in the Enhanced Oil Recovery (EOR). This involves injecting CO_2 into oil deposits. CO_2 reduces the viscosity of the oil and increases the pressure in the oil-bearing rock, which makes the exploitation easier [60]. This makes it possible to exploit deposits in which the natural pressure has decreased as a result of previous extraction. This extends the time of use of the deposits, the exploitation of which would no longer be profitable without this procedure. Similarly, in the case of natural gas, the Enhanced Gas Recovery (EGR) can be introduced [61]. Methods of this type have been used for more than 40 years in Canada and the USA. By using CO_2 obtained from the waste of the combustion process, the need for purchase is eliminated, which will reduce the price of the fossil fuel. Additionally, carbon dioxide can be successfully stored in depleted gas and oil deposits. CO_2 is also used in the coal gasification process. It is used as a gasifying agent and oxygen carrier. Carbon dioxide causes an increased degree of coal reaction and increases the share of CO_2 in the synthesis gas [62]. This increases the efficiency of the gasification process and reduces financial outlays.

Due to its high heat transfer coefficient, carbon dioxide can also be used in geothermal installations as a medium with much higher efficiency than classic water-steam systems [63]. CO_2 can also play a role in the process of increasing the pressure in the geothermal reservoir, which decreases during operation in installations where extracted water is not reinjected into the deposit after its heat has been removed. This solution also allows for combining the processes of obtaining geothermal energy and sequestration [64]. Power-to-gas-to-power technology has also been developed, which involves storing methane and carbon dioxide in underground tanks. This is a solution that allows for the storage of excess energy from renewable sources until it is needed. This energy is used in the process of electrolysis of water, thanks to which hydrogen is produced. Then, the CO_2 recovered and stored in the tank undergoes a chemical reaction with hydrogen, which results in the production of methane. Methane is stored in underground geological formations until it is extracted to the surface and burnt in heating plants or power plants. During its combustion, the released CO_2 is captured and re-collected in an underground reservoir, creating a closed carbon cycle [65]. CO_2 can also be used in the production of synthetic fuels. From the collected CO_2 and the obtained hydrogen, methanol, methyl alcohol, gasoline, and diesel fuel, formic acid can be obtained [66, 67]. They can also be hydrogen carriers, e.g. formic acid can be easily converted back to hydrogen. CO_2 is also used in the production of plastics, where it can replace petroleum-based raw materials, resulting in both environmental and economic benefits [68]. Urea can also be produced using CO_2. Urea is used in agriculture as a nitrogen fertiliser and for the production of chemicals such as resins [69]. Carbon dioxide is also used in the production of concrete. Thanks to the reaction of CO_2 with calcium hydroxide, the concrete obtained is more resistant to external factors and loads [70]. Carbon dioxide can be used in greenhouses to enrich the atmosphere, which translates into abundance of crops and crop efficiency. CO_2 is one of the most important factors necessary in the photosynthesis process. It can be used in algae cultivation as a factor accelerating their growth [71]. Algae are used in the production of biofuels, in the textile industry, in wastewater treatment plants, and fertilizers [72, 73]. Carbon dioxide can also be used in refrigeration systems, as a cooling agent in air conditioning systems, to cool food in cold stores, or in industrial processes. This has mainly advantages from the point of view of environmental protection, because it has a less negative impact than freons, and CO_2 is also not flammable. CO_2 increases the efficiency of the cooling process [74].

Another gaseous waste produced during coal combustion is nitrogen oxide NO_x. The largest amounts of NO and NO_2 are produced as a result of fuel oxidation and nitrogen contained in the air involved in the combustion process [75]. These gases are toxic and, unlike CO_2, have an additional negative impact on living organisms. NO affects the expansion of the ozone hole [76]. The reaction of nitrogen dioxide with water leads to the formation of nitric acid, which acidifies the soil, surface, and groundwater. N_2O is one of the greenhouse gases that leads to an increase in the temperature of the planet. N_2O is particularly dangerous due to its long duration in the atmosphere [77]. Nitrogen dioxide in contact with moisture and high sunlight leads to the formation of a Los Angeles-type smog. Ozone is produced, which can lead to breathing problems, eye and respiratory tract irritation, and animal and plant damage [78]. Due to CCT, such as nitrogen oxide selective membranes, they can be separated and utilised in a similar way to CO_2. One of the solutions is to use the obtained oxides to produce nitric acid [79]. Nitric acid has many applications, it is a very important raw material in the chemical industry, used not only for the production of fertilisers but also in the process of recovering REEs from fly ash. In the future, it can also be used in energy storage systems. However, at present, these are solutions at the stage of basic research.

During coal combustion, the sulphur compounds contained in it undergo the combustion process and become part of the exhaust gases. As a result of this reaction, sulphur dioxide and SO_3 are produced. SO_2 is a toxic gas with a harmful effect on living organisms. It can lead to diseases of the respiratory system and circulatory system. Reacting with water, SO_2 is transformed into sulfuric acid, leading to the formation of acid rain [80]. Acid rains acidify water, soils, and degrades plants.

SO_2 recovered from exhaust gases can be processed into sulfuric acid, which is widely used in the chemical, petrochemical, metallurgical, paper, pharmaceutical, textile industries, etc. [81] Sulfuric acid can also be used in a similar way to nitric acid in the recovery of rare earth elements from fly ash [82]. The reaction of SO_2 with calcium carbonate allows the production of building gypsum, while sulphites are used in the food industry for food preservation [83, 84].

References

1. Sobczyk EJ, Wota A, Kopacz M, Frączek J (2017) Clean coal technologies—a chance for Poland's energy security. Decision-making using AHP with benefits, opportunities, costs and risk analysis. Gospodarka Surowcami Mineralnymi 33(3):27–48
2. Pawlak-Kruczek H, Ostrycharczyk M, Czerep M, Baranowski M, Zgóra J (2015) Examinations of the process of hard coal and biomass blend combustion in OEA (oxygen enriched atmosphere). Energy 92:40–46
3. Mével R, Sabard J, Lei J, Chaumeix N (2016) Fundamental combustion properties of oxygen enriched hydrogen/air mixtures relevant to safety analysis: experimental and simulation study. Int J Hydrogen Energy 41(16):6905–6916
4. Liu Z, Zhang T (2018) Pilot and industrial demonstration of oxy-fuel combustion. In: Oxy-fuel combustion. Academic Press, pp 209–222
5. Callide oxyfuel project. CS energy
6. Vostrikov AA, Dubov DY, Psarov SA, Sokol MY (2007) Combustion of coal particles in H_2O/O_2 supercritical fluid. Ind Eng Chem Res 46(13):4710–4716
7. Advanced Ultra-Supercritical Technology. GE Steam Power. https://gevernova.com
8. Walton SM, He X, Zigler BT, Wooldridge MS (2007) An experimental investigation of the ignition properties of hydrogen and carbon monoxide mixtures for syngas turbine applications. Proc Combust Inst 31(2):3147–3154
9. Lee S (2007) Gasification of coal. In: Handbook of alternative fuel technologies, pp 26–76
10. Liu K, Cui Z, Fletcher TH (2009) Coal gasification. In: Hydrogen and syngas production and purification technologies, pp 156–218
11. Maxim V, Cormos CC, Cormos AM, Agachi S (2010) Mathematical modeling and simulation of gasification processes with carbon capture and storage (CCS) for energy vectors polygeneration. Comput Aided Chem Eng 28:697–702
12. Ariyapadi S, Shires P, Bhargava M, Ebbern D (2008) KBR'S transport gasifier (TRIG™)—an advanced gasification technology for SNG production from low-rank coals
13. Pohořelý M, Vosecký M, Hejdová P, Punčochář M, Skoblja S, Staf M, Svoboda K (2006) Gasification of coal and PET in fluidized bed reactor. Fuel 85(17–18):2458–2468
14. Porada S, Czerski G, Dziok T, Grzywacz P (2013) Coal gasification technologies and their suitability for the needs of energy and chemistry. Przegląd Górniczy 69(2):200–208
15. Sudiro M, Pellizzaro M, Bezzo F, Bertucco A (2010) Simulated moving bed technology applied to coal gasification. Chem Eng Res Des 88(4):465–475
16. Czerski G, Dziok T, Strugała A, Porada S (2014) Assessment of coal gasification technologies from the point of view of their suitability for the chemical industry. Chem Industry 93(8):1393–1400
17. Wang T, Stiegel GJ (eds) (2016) Integrated gasification combined cycle (IGCC) technologies. Woodhead Publishing
18. Brown KM (2012) In situ coal gasification: an emerging technology. J Am Soc Min Reclam 1(1)
19. Wiatowski M, Stańczyk K, Świądrowski J, Kapusta K, Cybulski K, Krause E, Smoliński A (2012) Semi-technical underground coal gasification (UCG) using the shaft method in experimental mine "Barbara." Fuel 99:170–179
20. Su F, Nakanowataru T, Itakura KI, Ohga K, Deguchi G (2013) Evaluation of structural changes in the coal specimen heating process and UCG model experiments for developing efficient UCG systems. Energies 6(5):2386–2406
21. Luo Y, Coertzen M, Dumble S (2009) Comparison of UCG cavity growth with CFD model predictions. In: Proceedings of the seventh international conference on CFD in the minerals and process industries (CSIRO), Melbourne, Australia, pp 9–11
22. Friedmann SJ, Upadhye R, Kong FM (2009) Prospects for underground coal gasification in carbon-constrained world. Energy Procedia 1(1):4551–4557
23. Yi T, Qin Y, Zhou Y, Wang L, Jin J, Zhou Z (2023) Research advances on the techno-economic evaluation of UCG projects. Coal Geol Explor 51(7):2

24. Czaja P (2014) Energy from coal obtained by gasification. Chemik 68(12)
25. Rosen MA, Reddy BV, Self SJ (2018) Underground coal gasification (UCG) modeling and analysis. In: Underground coal gasification and combustion, pp 329–362
26. Otoshi A, Sasaki K, Anggara F (2022) Screening of UCG chemical reactions and numerical simulation up-scaling of coal seam from laboratory models. Combustion Theor Modeling 26(1):25–49
27. Self SJ, Reddy BV, Rosen MA (2012) Review of underground coal gasification technologies and carbon capture. Int J Energy Environ Eng 3(1):1–8
28. Hilsenrath PE (1985) Coal-based synthetic fuels: the South African experience (synfuels, sasol, energy). Doctoral dissertation, The University of Texas at Austin
29. Dancuart LP, Steynberg AP (2007) Fischer-Tropsch based GTL technology: a new process? In: Studies in surface science and catalysis,163. Elsevier, 379–399
30. European Energy Exchange AG (EEX). Available online https://www.eex.com. Accessed on 2 Sept 2024
31. Polski CCS (2024) Polski CCS – Szansa na redukcję emisji CO_2. Polski Instytut Ekonomiczny. Available online https://pie.net.pl/wp-content/uploads/2024/07/Polski-CCS-co2.pdf. Accessed on 20 Sept 2024
32. van Alphen K, Hekkert MP, Turkenburg WC (2009) Comparing the development and deployment of carbon capture and storage technologies in Norway, the Netherlands, Australia, Canada and the United States—an innovation system perspective. Energy Procedia 1(1):4591–4599
33. Cook PJ (2017) CCS research development and deployment in a clean energy future: lessons from Australia over the past two decades. Engineering 3(4):477–484
34. Motowidlak T (2018) Poland's dilemmas in implementing the European Union's energy policy. Polityka Energetyczna Energy Policy J 21(1):5–20
35. Sowiżdżał A, Starczewska M, Papiernik B (2022) Future technology mix—enhanced geothermal system (EGS) and carbon capture, utilization, and storage (CCUS)—an overview of selected projects as an example for future investments in Poland. Energies 15(10):3505
36. Ahmadi M, Janakiram S, Dai Z, Ansaloni L, Deng L (2018) Performance of mixed matrix membranes containing porous two-dimensional (2D) and three-dimensional (3D) fillers for CO_2 separation. Membranes 8(3):50
37. Kusworo TD, Budiyono IAF, Mustafa A (2015) Fabrication and characterization of polyimide-CNTs hybrid membrane to enhance high performance CO_2 separation. Int J Sci Eng 8(2):115–119
38. Perez EV, Balkus KJ, Ferraris JP, Musselman IH (2009) Mixed-matrix membranes containing MOF-5 for gas separations. J Membr Sci 328(1–2):165–173
39. Ahmad J, Hagg MB (2013) Development of matrimid/zeolite 4A mixed matrix membranes using low boiling point solvent. Sep Purif Technol 115:190–197
40. Sridhar S, Smith B, Aminabhavi TM (2006) Modified poly (phenylene oxide) membranes for the separation of carbon dioxide from methane. J Membr Sci 280(1–2):202–209
41. Bershtein VA, Egorova LM, Yakushev PN, Georgoussis G, Kyritsis A, Pissis P, Sysel P, Brozova L (2002) Molecular dynamics in nanostructured polyimide-silica hybrid materials and their thermal stability. J Polym Sci Part B Polym Phys 40(10):1056–1063
42. Luebke D, Myers C, Pennline H (2006) Hybrid membranes for selective carbon dioxide separation from fuel gas. Energy Fuels 20(5):1906–1913
43. Vu DQ, Koros WJ, Miller SJ (2003) Mixed matrix membranes using carbon molecular sieves: I. Preparation and experimental results. J Membr Sci 211(2):311–334
44. Vinh-Thang H, Kaliaguine S (2013) Predictive models for mixed-matrix membrane performance: a review. Chem Rev 113(7):4980–5028
45. Sysel P, Minko E, Hauf M, Friess K, Hynek V, Vopicka O, Pilnacek K (2011) Mixed matrix membranes based on hyperbranched polyimide and mesoporous silica for gas separation. Desalin Water Treat 34(1–3):211–215
46. Suzuki T, Yamada Y (2005) Physical and gas transport properties of novel hyperbranched polyimide/silica hybrid membranes. Polym Bull 53:139–146

47. Krystl V, Hradil J, Bernauer B, Kocirik M (2001) Heterogeneous membranes based on zeolites for separation of small molecules. React Funct Polym 48:129–139
48. García Giménez R, Vigil de la Villa R, Frías M (2016) From coal-mining waste to construction material: a study of its mineral phases. Environ Earth Sci 75:1–8
49. Yen HW, Ho SH, Chen CY, Chang JS (2015) CO_2, NO_x and SO_x removal from flue gas via microalgae cultivation: a critical review. Biotechnol J 10(6):829–839
50. Statystyka Polska. Available online from https://stat.gov.pl/. Accessed 15 Sept 2024
51. Mathapati M, Amate K, Prasad CD, Jayavardhana ML, Raju TH (2022) A review on fly ash utilization. Mater Today Proc 50:1535–1540
52. Wdowin M, Franus W (2014) Analysis of fly ash for the purpose of obtaining rare earth elements. Energy Policy J 17(3):369–380
53. Czech T, Jaworek A, Marchewicz A, Krupa A, Sobczyk AT (2017) Heavy metal content in fly ash. In: XXIV international conference ashes from energy
54. Solovyova VM, Ilinova AA, Cherepovitsyn AE (2020) Strategic forecasting of REE mining projects development in Russian arctic. In: Advances in raw material industries for sustainable development goals. CRC Press, pp 456–464
55. Buechler DT, Zyaykina NN, Spencer CA, Lawson E, Ploss NM, Hua I (2020) Comprehensive elemental analysis of consumer electronic devices: rare earth, precious, and critical elements. Waste Manage 103:67–75
56. Krishnamurthy P (2020) Rare metal (RM) and rare earth element (REE) resources: world scenario with special reference to India. J Geol Soc India 95:465–474
57. Pie P (2023) 98% of EU demand for rare earth elements is met by China. Available online https://pie.net.pl/en/98-of-eu-demand-for-rare-earth-elements-is-met-by-china/. Accessed 20 Sept 2024
58. Całus-Moszko J, Białecka B (2013) Analysis of the possibilities of obtaining rare earth elements from hard coal and fly ash from power plants. Gospodarka Surowcami Mineralnymi Min Resour Manage 29(1):67–80
59. Latacz A (2017) Determination of rare earth elements in coals and ashes. Works Institute Ferrous Metallurgy, 69
60. Lakatos I, Lakatos-Szabo J (2008) Global oil demand and role of chemical EOR methods in the 21st century. Int J Oil Gas Coal Technol 1(1–2):46–64
61. Fanchi JR (2005) Principles of applied reservoir simulation. Elsevier
62. Chmielniak T, Ściążko M, Sobolewski A, Tomaszewicz G, Popowicz J (2012) Coal gasification using CO_2 as a method for improving emission factors and process efficiency. Energy Policy 15(4):125–138
63. Wu Y, Li P (2020) The potential of coupled carbon storage and geothermal extraction in a CO_2-enhanced geothermal system: a review. Geothermal Energy 8(1):19
64. Wojnicki M, Kuśnierczyk J, Szuflita S, Warnecki M (2022) Integration of geothermal energy with mineral CO_2 sequestration. Nafta-Gaz 78
65. Kühn M, Streibel M, Nakaten N, Kempka T (2014) Integrated underground gas storage of CO_2 and CH_4 to decarbonize the "power-to-gas-to-gas-to-power" technology. Energy Procedia 59:9–15
66. Ganesh I (2014) Conversion of carbon dioxide into methanol–a potential liquid fuel: fundamental challenges and opportunities (a review). Renew Sustain Energy Rev 31:221–257
67. Zhang C, Hao X, Wang J, Ding X, Zhong Y, Jiang Y, Xiong Y (2024) Concentrated formic acid from CO_2 electrolysis for directly driving fuel cell. Angew Chem Int Ed 63(13):e202317628
68. van Heek J, Arning K, Ziefle M (2017) Reduce, reuse, recycle: acceptance of CO_2-utilization for plastic products. Energy Policy 105:53–66
69. Yildirim E, Guvenc I, Turan M, Karatas A (2007) Effect of foliar urea application on quality, growth, mineral uptake and yield of broccoli (*Brassica oleracea* L., var. italica). Plant Soil Environ 53(3):120
70. Tam VW, Butera A, Le KN (2020) Microstructure and chemical properties for CO_2 concrete. Constr Build Mater 262:120584

71. Iglina T, Iglin P, Pashchenko D (2022) Industrial CO_2 capture by algae: a review and recent advances. Sustainability 14(7):3801
72. Demirbas A (2010) Use of algae as biofuel sources. Energy Convers Manage 51(12):2738–2749
73. Mehra R, Bhushan S, Gill BS, Rehman WU, Bast F (2018) Algae-based composites and their applications. In: Biocomposites. Jenny Stanford Publishing, pp 163–179
74. Dilshad S, Kalair AR, Khan N (2020) Review of carbon dioxide (CO_2) based heating and cooling technologies: past, present, and future outlook. Int J Energy Res 44(3):1408–1463
75. Xu X, Chen C, Qi H, He R, You C, Xiang G (2000) Development of coal combustion pollution control for SO_2 and NO_x in China. Fuel Process Technol 62(2–3):153–160
76. Abbasi SA, Abbasi T, Abbasi SA, Abbasi T (2017) The ozone hole. In: Ozone hole: past, present, future, pp 13–35
77. Fluckiger J, Dallenbach A, Blunier T, Stauffer B, Stocker TF, Raynaud D, Barnola JM (1999) Variations in atmospheric N_2O concentration during abrupt climatic changes. Science 285(5425):227–230
78. Rani B, Singh U, Chuhan AK, Sharma D, Maheshwari R (2011) Photochemical smog pollution and its mitigation measures. J Adv Sci Res 2(04):28–33
79. Radsak D (2017) Reduction of nitrogen oxides emissions in power boilers as a necessity to meet European emission standards. Poznan Univ Technol Acad J Electr Eng
80. Gimeno L, Marín E, Del Teso T, Bourhim S (2001) Effective reduction of SO_2 emissions on the effect of acid rain on ecosystems. Sci Total Environ 275(1–3):63–70
81. Roy P, Sardar A (2015) SO_2 emission control and finding a way out to produce sulfuric acid from industrial SO_2 emission. J Chem Eng Process Technol 6(2):1000230
82. Tayar SP, Palmieri MC, Bevilaqua D (2022) Sulfuric acid bioproduction and its application in earth rare extraction from phosphogypsum. Miner Eng 185:107662
83. Ennaciri Y, Zdah I, El Alaoui-Belghiti H, Bettach M (2020) Characterization and purification of waste phosphogypsum to make it suitable for use in the plaster and cement industry. Chem Eng Commun 207(3):382–392
84. Quattrucci E, Masci V (1992) Nutritional aspects of food preservatives. Food Addit Contam 9(5):515–525

Chapter 4
Coal Energy Technologies and Renewable Energy Sources

Abstract This chapter presents the benefits of using hybrid energy systems based on clean coal technologies and renewable energy sources. The idea of an energy-chemical cluster and a financial analysis of the profitability of building a cluster were presented. The connections and dependencies of the units included in the cluster were characterized and the results of the spatial analysis for an example cluster were presented.

Keywords Energy-chemical clusters · Renewable energy sources · Financial analysis

Coal-based energy technologies and renewable sources of energy are usually perceived as two separate, mutually exclusive sources of energy. It is assumed that RES should completely take over the role of fossil fuels as soon as possible. Because of this, the goal of energy transition is to be achieved, i.e. reducing emissions of harmful substances that are created during the combustion of fossil fuels. However, RES and CCT should rather be perceived as two mutually supporting elements of a coherent and zero-emission energy system. Thanks to their joint use, a synergistic effect can be achieved, which will ensure the effective implementation of the goals specified in the Paris Agreement, the Fit for 55 Package, set by the SDGs, and the European Green Deal.

4.1 Hybrid Energy Solutions

Coal, thanks to CCT technologies, can be used as a bridge enabling the transition from traditional energy systems to sustainable and zero-emission systems. This is particularly justified in the case of countries with their own coal deposits, where it has been the main source of energy in most cases to this day. However, to continue to use the benefits of coal, a new approach to energy production is necessary, both in terms of the cleanliness of the combustion process, stabilisation of the energy system,

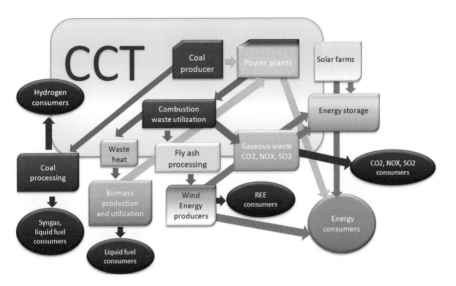

Fig. 4.1 Energy-chemical cluster diagram, *source* own elaboration

effective combustion, effective and efficient management of the energy produced, and also from an economic point of view. One of the possible solutions is to build energy-chemical clusters (Fig. 4.1).

The concept of clusters was initiated by M. Porter in the 1990s [1]. He characterised clusters as geographic concentrations of related companies, specialised suppliers, service providers, companies from related sectors, and associated institutions, such as universities and industry associations, which compete and cooperate with each other [2, 3]. According to Porter, creating clusters has a huge impact on innovation, development, productivity, and financial effects achieved by the entities that are part of it. Integrated coal combustion systems, in addition to mines, power plants, and heating plants, should include research units whose task will be to work on clean coal technologies, technological and organisational solutions that will affect the economic and environmental results achieved by the clusters. These technologies will enable the optimisation of the combustion process and the separation and processing of waste generated during coal combustion, i.e., mainly REEs, CO_2, NO_x, SO_x, into valuable raw materials. In addition, the cluster should also include chemical plants, where the process of obtaining the substances and metals and their proper preparation in accordance with the expectations of recipients (e.g. purification, creation of appropriate concentrates in the required chemical form) will be carried out. Some of the substances produced may also be used within the cluster. The cluster operates in accordance with the assumptions of the circular economy. In addition to technological solutions that will enable and streamline the processes of extraction, processing fuel and its waste, an economic effect can be achieved by using only organisational solutions in an appropriate manner, coordinating the work of the units included in the cluster. This mainly concerns the optimisation of the supply of raw materials, energy, heat, and the use of just-in-time deliveries. The reduction of the operating costs can also be achieved by designing the shortest supply routes at a given time. Energy efficiency and productivity that will be achieved within the cluster can also be seen as an additional source of energy.

The cluster can be described by the following equation:

$$K = \{P, S, A, I, R\}$$

where

P   potential, indicates the cluster resources that are already available and can be used in the future,
S   synergy—manifesting itself in additional benefits obtained thanks to cooperation within the cluster,
A   cluster actors, environments associated within the cluster,
I   cluster idea, vision, and cluster goals,
R   relationship—defines the connections between cluster actors, describing its entrepreneurial culture.

Energy and chemical clusters are, therefore, structures that include appropriately interconnected and organised production, processing and logistics units connected in order to achieve competitive advantage and environmental neutrality. An important aspect that allows the proper functioning of the cluster is the flow of knowledge and information within the cluster. The creation of a cluster will provide the possibility of obtaining unwavering supplies of raw materials and energy, facilitate the meeting of the expectations of recipients, optimal allocation of resources, reduction of operating costs, and reduction of the level of operational risk. The functioning of units in the cluster facilitates the achievement of the feedback effect in the relations between its individual objects.

The distribution chain of products and semi-finished products within the energy-chemical cluster is shown in Fig. 4.2.

Factors that stimulate the construction of a cluster are primarily the spatial concentration of key facilities, access to research units, knowledge, workforce with appropriate qualifications, communication network, as well as support from the state, local government, and appropriate legal regulations. In countries where coal has been extracted for generations, such as Poland or China, finding a suitable location is easier. In the case of Poland, this is the Silesian Voivodeship, for which the authors conducted research on the construction of the clusters. Using geographic information systems and clustering tools, the facilities selected for analysis were divided into three groups (clusters) consisting of homogeneous facilities. During the analysis, the geographical coordinates of the facilities were taken into account. Additionally, it was assumed that the cluster should include facilities necessary to ensure its full functionality, i.e. a fuel producer, power plant, chemical plant, solar farms, wind farms, production and biomass units, etc. The next step is to properly organise the workflow within the created clusters. First, it is necessary to determine the level of

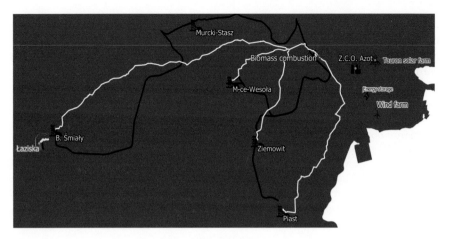

Fig. 4.2 Cluster spatial model, *source* own elaboration

demand for energy from each of the available sources. Due to the specific nature of the underground hard coal mining process, annual and monthly forecasts of fuel production volumes were made.

Forecasts will allow planning the volume of extraction in a given month in a given mine, which is part of the cluster. This will allow for the determination and ensuring of the appropriate level of production factors necessary to extract the planned amount of coal. Then, the coal must be transported, for example, to a power plant. Within the cluster, it is possible to develop the shortest route for transporting coal with specific quality parameters. The surplus coal production can be made available and sold to external recipients outside the cluster (e.g. export). In Poland, the annual costs of transporting coal within the country amount to several billion USD per year. The use of GIS system algorithms will allow indicating a supplier of coal with the required parameters and in the desired quantity. Polish coal is transported by rail and road. For the cluster presented, it was determined that the use of road transport, assuming the rates are identical, will reduce transport costs by about 3%, which, gives savings of USD 2 million per year and at least USD 30 million nationwide.

After the coal is extracted, it should be cleaned and enriched in a processing plant. Then, after it is transported to the first recipient in the distribution chain, it is processed into electricity or heat. This processing takes place in units equipped with appropriate CCTs. These can be, for example, supercritical power plants and membrane modules, which allow for the separation of the gas mixture and the acquisition of appropriate concentrates. One of the biggest challenges related to CCS technology is the transport of the recovered gas to the place of storage. Integrating units within a cluster that will use the obtained gas will eliminate this logistical and financial problem. CO_2 can be utilised in a unit producing biomass based on algae. The cultivation process requires appropriate thermal conditions, so to eliminate this problem, waste heat from the energy generation process in power plants can be used. The resulting algae can be co-combusted with coal, or can be processed into liquid fuels. The coal extracted within the cluster can also be gasified or converted into liquid fuel. Obtained syngas can be used to generate electricity, heat, synthetic natural gas, and hydrogen. Hydrogen is seen as a gas that can play a key role in the energy transition process. It can be used in power generation, heavy industry, energy storage, and transport. Solar and wind farms should also be located in the cluster area. The energy obtained from them should form the basis of the energy system. When, due to weather conditions and time of day, RES production ceases or is insufficient, the system must be supported by energy obtained from burning coal. The excess RES energy, in turn, should be stored. The cluster should contain energy storage facilities in which the obtained CO_2 and hydrogen will be used. This will again eliminate the problem of gas transport, which is particularly beneficial in the case of hydrogen. Solid waste from the coal combustion process, i.e. primarily fly ash, can be processed, e.g. using chemical extraction and membrane modules to obtain rare earth elements. Then, these are delivered to recipients, including wind technology manufacturers.

4.2 Energy-Chemical Clusters Profitability Analysis

Chemical substances obtained from waste will constitute an added value, and the funds obtained can be used to support the development of new technologies, research, and investments in CCT installations. The results of the financial analysis are presented in Table 4.1. It has been calculated that the production of 1 Mg of algae per year requires the consumption of about 2 Mg of CO_2 [4]. In Poland, the energy sector emits about 305 million Mg of CO_2 per year, which means that it will be possible to produce 171 million Mg of algae per year. Assuming that the cost of 1 Mg of CO_2 is at least USD 100 [5], this would bring savings of about USD 31 billion per year. The cost of emitting a ton in the ETS system is about USD 73 [6, 7], so CO_2 management gives savings of USD 22 billion per year.

Table 4.1 Results of the financial analysis of the CO_2 acquisition and management process, algae production

	Factor	Value
1	CO_2 emission, mil Mg	305
2	Algae production, mil Mg	171
3	CO_2 purchase cost, bn USD	31
4	CO_2 emissions trading cost, bn USD	22
5	Cost of CO_2 acquisition, membrane techniques, bn USD	15
7	Algae production cost, mil USD	115
8	Total savings, mil USD (3 + 4)	53
9	Income, bn USD	206
10	Profit, bn USD (9–5-7 + 8)	243

Savings related to the management of the extracted CO_2 amount to about USD 53 bn. Assuming that the cost of producing one Mg of algae is about USD 600 [8], it would amount to USD 115 bn per year. The average selling price of Mg algae is USD 1300 [9], therefore the revenues obtained would be USD 206 bn, while the profit reduced by production costs is USD 243 bn.

Assuming that the average cost of CO_2 separation using membrane techniques is 50 USD/Mg [10], it amounts to USD 15 billion per year. Taking into account the CO_2 values obtained in Table 4.2, savings related to emission trading and the cost of using CCT, the annual profit obtained thanks to CO_2 extraction alone amounts to about 38 billion dollars.

An analogous analysis was carried out for solid waste from the coal combustion process, i.e., fly ash. The estimation results are presented in Table 4.3.

In Poland, approximately 10 million Mg of fly ash is produced annually [11]. Taking into account its composition, one ton of Polish ash can produce REEs worth approximately USD 23,000 [12, 13]. It has been assumed that the operating costs of REEs recovery are approximately USD 15,000/Mg [14], which, taking into account the amount of ash, gives a total operating cost of USD 180 billion. Annual profit is USD 80 billion.

Energy losses produced by photovoltaic cells and wind turbines can amount to about 10–20% per year [15]. It was verified that the construction of energy storage facilities that would be able to store it is economically justified. The losses were assumed to 15% of the annual production of renewable energy sources. The value of energy that could be stored in the storage facilities and the cost of their construction were calculated, as presented in Table 4.4. The profit obtained in one year amounts to USD 0.8 billion. Therefore, the cost of building energy storage facilities would

Table 4.2 Results of the financial analysis of the CO_2 acquisition and management process

	Factor	Value
1	CO_2 value, USD bn	31
2	CO_2 emissions trading cost, bn USD	22
3	Cost of CO_2 acquisition, membrane techniques, USD bn	15
4	Income, bn USD (1 + 2)	53
5	Profit, bn USD (4–3)	38

Table 4.3 Economic analysis of the profitability of REEs recovery in Poland

Factor	Value
Opex, bn USD/year	150
Amount of ash produced annually in Poland, mil Mg	10
REEs value, USD/Mg of ash	23,000
REEs value (annual production), bn USD	230
Profit, bn USD	80

Table 4.4 Financial analysis of the construction of energy storage facilities

Factor	Value
Energy losses per year, EJ	0.06
Energy value, bn USD [16]	2.9
Cost of building an energy storage facility, USD bn	2.1
Profit, bn USD	0.8

pay off in the first year of their use, and in the following years the profits would be higher.

The presented idea of an energy-chemical cluster may constitute a solution in the situation of growing demand for energy, in the era of energy transition, and in connection with the necessity of decarbonisation of energy systems. It was presented on the example of Poland, as the most representative country in the European Union. Analogous solutions adapted to local conditions can be applied wherever there is a need to maintain coal or other fossil fuels in the energy mix, while eliminating environmental problems related to the combustion of these fuels and applying a solution that takes into account the circular economy. Building a cluster will enable optimisation of processes conducted in facilities that belong to it. An organisational solution of this type provides many benefits, both financial, such as reducing production and transport costs, eliminating the costs of lost opportunities, optimising employment, and just-in-time production. Organisational solutions are a cost-free form of optimising the functioning of individual cluster units. However, the most important feature of clusters is the use of clean coal technologies within them, which will require the financing of related investments in the initial phase. However, the possibility of using the chemicals obtained through them will constitute a source of financing for the operating costs of CCTs, their further development, and research conducted on their development. The financial analysis has shown that the funds obtained on the basis of waste from the incineration process can constitute a source of financing for CCTs.

The proposed CCT i.e. membrane techniques, are characterised by a number of features that distinguish them from other technological solutions. Membrane techniques can be used within the framework of existing energy infrastructure and are low energy.

References

1. Kuah AT (2002) Cluster theory and practice: advantages for the small business locating in a vibrant cluster. J Res Mark Entrep 4(3):206–228
2. Porter ME (2000) Location, competition, and economic development: local clusters in a global economy. Econ Dev Q 14(1):15–34
3. Porter ME (2009) On competition. Harvard Business School Publishing, Boston
4. Sundaramurthy S, Premalatha M (2011) An overview of CO_2 mitigation using algae cultivation technology. Int J Chem Res 3:110–117. https://doi.org/10.9735/0975-3699.3.3.110-117

5. IMARC Group. Carbon dioxide pricing report. Available online https://www.imarcgroup.com/carbon-dioxide-pricing-report. Accessed on 4 Sept 2024
6. Instrat. EU ETS CO_2 emission allowance prices. Available online https://www.instrat.pl/eu-ets-prices/. Accessed on 6 Sept 2024
7. Rybak A, Rybak A (2021) Methods of ensuring energy security with the use of hard coal—the case of Poland. Energies 14(18):5609. https://doi.org/10.3390/en14185609
8. Algae Biomass Organization. The next big step for algae and sustainable aviation fuels. Available online https://algaebiomass.org/blog/11551/the-next-big-step-for-algae-and-sustainable-aviation-fuels/. Accessed on 8 Sept 2024
9. Davis R, Markham J, Kinchin C, Grundl N, Tan ECD. Process design and economics for the production of algal biomass: algal biomass production in open pond systems and processing through dewatering for downstream conversion. Technical report NREL/TP-5100-64772. National Renewable Energy Laboratory
10. Adhikari B, Orme CJ, Stetson C, Klaehn JR (2023) Techno-economic analysis of carbon dioxide capture from low concentration sources using membranes. Chem Eng J 474:145876
11. GUS. Energy in 2023. Available online https://stat.gov.pl/energy_2023.pdf. Accessed on 10 Sept 2024
12. Wdowin M, Franus W (2014) Analysis of ash volatile at an angle to get from them elements lands rare. Energy Policy J 17(3):369–380
13. Institut für Seltene Erden und Metalle. Metal prices—high purity metals prices. Available online https://en.institut-seltene-erden.de/our-service-2/Metal-prices/high-purity-metals-prices/. Accessed on 12 Sept 2024
14. Chalmers I, MacDonald A (2017) Economics of rare earth projects (Alkane Resources LTD.). In: Materials science & technology (MS&T) conference, Pittsburgh, Pennsylvania, USA
15. Teraz Środowisko. Redysponowanie nawet 1 TWh utraconej zielonej energii w 2024. Available online https://www.teraz-srodowisko.pl/aktualnosci/redysponowanie-nawet-1-twh-utraconej-zielonej-energii-w-2024-15323.html. Accessed on 14 Sept 2024
16. Optimal Energy. Cena prądu dla firm w Polsce. Available online https://optimalenergy.pl/cena-pradu/#:~:text=W%20przypadku%20firm%2C%20ceny%20pr%C4%85du,dostawc%C3%B3w%20i%20region%C3%B3w%20w%20Poland. Accessed on 16 Sept 2024

Chapter 5
Clean Coal Technologies and Energy Security

Abstract This chapter discusses the concept of energy security, its definitions, and the results of the analysis of the energy security indicator. The impact of renewable energy sources and clean coal technologies on the energy security indicator was discussed. The results of the analysis of the impact of the total and sudden elimination of coal on energy security measures were also presented.

Keywords Energy security · Energy security index · REE

Events that have occurred in recent years have seriously undermined the energy security of countries around the world. First, the pandemic and then the war in Ukraine had a direct impact on access to energy resources. The decrease in energy demand during the pandemic and the change in energy consumption patterns led to a collapse in fuel prices in global markets. However, the war that began in February 2022 and overlapped with the pandemic caused a rebound in energy demand and prices in the opposite direction. First of all, problems with the supply of natural gas, oil, and coal from Russia occurred, as a result of sanctions on the import of energy resources from that country. This also affected the prices of raw materials, which significantly have increased inflation, and led to an economic crisis in many countries. It was once again noted that reliance on imported energy carriers from politically unstable regions can lead to a threat to energy security and therefore also military security. The definition of energy security first appeared in the 1970s and was formulated after the oil crises, which at that time turned oil into a political weapon and led to several-fold increases in fuel prices, ultimately leading Western countries to an energy crisis [1]. The definition of energy security at that time was based mainly on the stability of raw material supplies and their prices [2, 3]. Currently, the definition is evolving, taking into account an increasingly broader spectrum of factors that determine the energy comfort of a given country. It primarily included the impact of energy on the natural environment and human health and life. Energy security consists of factors classified into four groups, i.e. accessibility, availability, acceptability, affordability called 4A [4]. This definition should soon be expanded to include issues related to energy transition so that it can be considered fully complete.

© The Author(s), under exclusive license to Springer Nature Switzerland AG 2025
A. Rybak, *The Role of Clean Coal Technologies in Energy Transformation and Energy Security*, SpringerBriefs in Energy, https://doi.org/10.1007/978-3-031-80652-0_5

One of the most well-known definitions is the definition proposed by the IEA, where energy security is described as the uninterrupted availability of energy sources at an affordable price [5]. Due to the ongoing changes in the energy systems of countries around the world, this definition should be modified in accordance with the changes that are taking place in the structure of energy production. The current definitions have been developed mainly for fossil fuels, and in accordance with them, indicators for measuring energy security. These indicators include one or more factors classified in category 4A, such as reserves, macroeconomic indicators, diversification indicators, energy balance indicators, gas flow indicators, infrastructure indicators, sectoral indicators, import risk indicators, crisis indicators, [6], carriers, primary energy sources, end- use sector, resilience indicators, sovereignty indicators, robustness indicators, compound indicators [7], energy demand, energy carrier resources, energy carrier prices, energy prices, surplus energy production and demand, energy efficiency, share of RES in the energy mix, etc. [8]. RES are perceived in the context of energy security as a reliable and inexhaustible source of energy that does not require import. However, this issue should be viewed in a broader context, also taking into account the threats to energy security that excessive reliance on renewable energy sources may pose. Also in this case, the principle of diversification of energy sources should be remembered. Increasing reliance on RES may involve several traps that are usually unnoticed or omitted. First, the following aspects of energy security should be noted in this context:

- Availability—renewable energy provides energy produced within the geographical area of a given country but requires the use of energy storage, otherwise it destabilises the energy system.
- Accessibility—renewable energy is theoretically available in every country, and its potential is unlimited. However, it should be remembered that, in addition to dependence on weather conditions, access to renewable energy also requires the use of appropriate technology. In most cases, this technology is imported mainly from China. If the supply chain of renewable energy technologies is disrupted, access to it will be limited. On the other hand, the production of photovoltaic panels and wind turbines requires access to appropriate raw materials. In the case of windmills, these are also rare earth elements, which are obtained almost exclusively from China.

In connection with this issues, an energy security measure was built that takes into account the access to rare earth elements [8]. The authors carried a research was on the example of the EU and Poland. The measure took into account the demand for REE elements. These were the prices of rare earth elements, the level of import and resources of elements, and the MVIS Global Rare Earth/Strategic Metals Index (MVREMX). SEM (synthetic energy security measure) and a simulation of changes in the indicator level was carried out, i.e., its value was calculated with and without taking into account the impact of REEs on energy security. It was noticed that taking into account the impact of REEs on the amount of energy security strongly affects the level of the SEM indicator. For Poland it decreased by 7%, for the entire European Union by 12%. As part of the simulation, the number of CCT patents was introduced to the SEM indicator. This influenced the increase of the SEM indicator. In the case of Poland it increased by 6%, and for the entire EU by 8%.

The SEM indicator was positively influenced, especially in the case of Poland, by hard coal resources and the surplus of energy production over its consumption. In the case of the EU, it was energy productivity and the level of CO_2 emissions. Factors negatively influencing the energy security indicator in the case of Poland, in addition to dependence on REEs imported from a single source, were identified as dependence on crude oil and natural gas and electricity imports, as well as the level of CO_2 emissions. On the scale of the entire European Union, the main threat is dependence on imports of energy resources and REEs.

The level of the SEM indicator, and consequently the energy security, can be raised by eliminating the weaknesses of energy systems. In turn, each of the weaknesses of energy systems in the EU can be eliminated by using clean coal technologies.

The value of the energy security indicator SEM can be increased in the case of Poland by ensuring a decrease in CO_2 emissions. After intensive decreases in the 1990s, emission reduction stabilised due to the lack of implementation of subsequent technical solutions and methods for its further decrease. The clean coal technologies mentioned in Chap. 3 are an excellent solution to this problem. CCT will also provide the possibility of diversifying primary energy by the possibility of maintaining an additional components in the energy mix. They also influence positively energy efficiency. Furthermore, for countries that have their own resources of this fossil fuel, the level of the import dependency indicator will be reduced. The price of renewable energy is currently often comparable to that of energy obtained from fossil fuels. This was mainly caused by the decrease in the costs of RES technologies, the increases in the prices of fossil fuels, the costs related to the emission trading system, and the financial support mechanisms for renewable energy. However, the use of solutions as in the proposed energy-chemical clusters could significantly reduce the prices of cogenerated energy based on coal (or other fossil fuel) and renewable energy sources (Chap. 4). In turn, the surplus of energy production over demand can be achieved by using the energy storage facilities included in the cluster. Obtaining REEs from fly ash will provide, first, the possibility of developing domestic wind turbine technology, supporting neighbouring countries with supplies of these critical raw materials, and also access to funds enabling further expansion and modernisation of the energy system.

Coal resources in Poland would suffice for about 200 years if used rationally. They will certainly not be exploited for such a long period of time. Considering the problem of energy security on a European Community scale and taking into account the fact that most countries in the EU do not have domestic fossil fuel resources, Poland could be an energy reservoir stabilising the energy systems of neighbouring countries. Of course, this would require a huge amount of work and financial resources for the systems of EU countries to cooperate to such an extent, but it would provide the possibility of optimal use of the energy potential of fossil fuels and renewable energy, of which countries more privileged in terms of climate produce much more and more efficiently. A similar solution can also be applied in other countries, using the potential of coal regions such as the Powder River Basin (USA) or Shanxi Province (China).

In turn, substituting coal with other fossil fuels is pointless. First, due to the chemical composition of natural gas and crude oil, which also causes them to emit greenhouse gases. Secondly, the analysis of the level of security indicators carried out on the example of Poland showed that eliminating coal and replacing it with natural gas or crude oil will lead to an increase in import dependence indicators. In the case of gas, this was an increase of 18% (to 94%), in the case of oil, the indicator increased from 98 to 99%. The Herfindahl–Hirschman Diversification Index (HHI) calculated for substitution with gas increased by 60%, for oil by 100%. This means a monopoly on the energy market. In turn, the Stirling Diversification Index for gas decreased by 30%, for oil by 45%. The energy self-sufficiency indicator for gas decreased by 16% and amounted to 5%, for oil it decreased by 11% and amounted to only 1% [9]. The simulation carried out showed that the removal of coal from the mixes of countries strongly dependent on its presence will have a negative impact on all energy security indicators. Until recently, natural gas was perceived in Europe as a blue fuel of the transition period, which will allow a smooth transition to zero-emission energy. However, tightening emission standards and the armed conflict in Ukraine have changed the view on natural gas. Therefore, it is necessary to ensure a smooth, gradual transition from energy systems based on fossil fuels to systems based on another energy source. Currently, all the hopes of world decision makers are placed in renewable energy sources. However, to maintain energy security at least at the current level, RES will require support. It should be noted that clean coal technologies are a solution that is mostly ready for use, with a recognized technology that has been developed for years and can cooperate with the existing energy system. CCTs are a ready solution to problems that will certainly appear during the energy transition, in every country that will work on transforming its energy system, whether it concerns coal, oil, or natural gas. The social aspect of the changes taking place should also be taken into account. Clean coal technologies can not only facilitate the preservation of current jobs, but also generate new workplaces. For coal regions, this is an important aspect that will allow avoiding sudden and unfavourable economic changes.

In summary, clean coal technologies can contribute to increased energy security in terms of access to energy through the diversification of energy carriers, in the environmental aspect they will eliminate the emission of harmful substances. They will also ensure the stabilisation of the energy system and energy supplies. They can also provide access to rare earth elements, which are key to the development of wind energy and electric transport.

References

1. Yergin D (1988) Energy security in the 1990s. Foreign Affairs 67(1):110–132
2. Lubell H (1961) Security of supply and energy policy in Western Europe. World Politics 13(3):400–422
3. Colglazier EW, Deese DA (1983) Energy and security in the 1980s. Ann Rev Energy 8(1):415
4. Asia Pacific Energy Research Center (APERC) (2007) A quest for energy security in the 21st century
5. IEA. Ensuring energy security. Available online https://www.iea.org/areas-of-work/ensuring-energy-security. Accessed on 7 Sept 2024
6. European Commission Joint Research Center Institute for Energy. Energy Security Indicators. Institute for Energy
7. Jewell J, Cherp A, Riahi K (2012) Energy security indicators for use in integrated assessment models. LIMITS Project, Project 282846, Deliverable 4.1
8. Rybak A, Rybak A, Kolev SD (2023) A synthetic measure of energy security taking into account the influence of earth rare metals. The case of Poland. Energy Rep 10:1474–1484
9. Rybak A (2020) Poland's energy security and coal position in the country's energy mix. Silesian University of Technology, Monograph 865, Gliwice

Chapter 6
Prospects for the Development of Clean Coal Technologies

Abstract This chapter considers the potential of CCT as a factor that eliminates the weak points of decarbonization policy. New directions of CCT development were characterized, focusing on hybrid membrane systems. The results of the SWOT analysis for clean coal technologies were also presented. The analysis allowed identification of the strengths, weaknesses of CCT and the opportunities and threats to their development located in the macroenvironment of energy systems. The possible course of the future development of clean coal technologies was also outlined.

Keywords CCT · SWOT analysis · MWCNTs membranes

The future of clean coal technologies will be decided in the near future. It will depend to the greatest extent on the policy adopted in a given region. The future of CCTs may vary depending on the decarbonisation path chosen by a given country or a community of countries. The potential of CCTs as a factor to eliminate the weak points of decarbonisation policies is enormous. The technologies are ready for implementation, and organisational solutions have been developed. Energy-chemical clusters are mostly based on existing infrastructure, and the potential profits that can be achieved within clusters show that they can be a self-financing structure that brings significant added value to the energy production process. CCTs should be seen as a tool to support the target zero-emission energy systems based on renewable energy sources or other clean energy sources. Therefore, their capabilities should be appreciated especially in the ongoing transition period. The importance of CCTs and their contribution to the transformation of energy systems will probably be different for developed or developing countries, but it should also depend on the structure of the energy mix of a given country. In countries that still rely heavily on coal, such as China, India, Indonesia, or Poland, CCTs should occupy a key position in sustainable energy development strategies. China, where energy demand is growing rapidly every year, is working to develop a significant share of renewable energy in the energy mix, while using coal. Therefore, the interest in CCTs in this country is constantly growing. The UN also perceives CCTs as one of the important factors of the sustainable development strategy. In the near future, however, it will be important

© The Author(s), under exclusive license to Springer Nature Switzerland AG 2025
A. Rybak, *The Role of Clean Coal Technologies in Energy Transformation and Energy Security*, SpringerBriefs in Energy, https://doi.org/10.1007/978-3-031-80652-0_6

to refine existing CCTs, as well as to develop new technologies that will eliminate the disadvantages of already existing solutions.

6.1 New Directions of Research in the Field of Clean Coal Technologies

Research conducted on CCT currently focusses mainly on its effectiveness and improving the parameters achieved. Research is also being conducted on new technologies that enable the processing of waste from the energy production process into valuable chemicals. In addition to the solutions described in Chaps. 3 and 4, research on photocatalytic CO_2 reduction is worth noting [1]. This process would allow the conversion of anthropogenic CO_2 with the help of solar energy to synthetic fuels.

The authors, in turn, are working on the use of membrane techniques to separate CO_2 from exhaust gases and recover selected rare earths elements from extracts of coal fly ashes.

As it was mentioned before the development of civilization and progressive industrialization requires increasing demand for energy, which in turn is related to the increase in the combustion of fossil fuels and greenhouse gas emissions, especially carbon dioxide [2]. In the last century, the increase in energy production from fossil fuels, the development of transport and the production of raw materials (steel, cement, etc.) were responsible for the increase in greenhouse gas emissions, mainly CO_2. Carbon dioxide emission constitutes more than half of the emissions of all greenhouse gases and unfortunately is responsible for global warming, growing sea levels and serious climate change. Therefore, currently all activities should be aimed at reducing greenhouse gas emissions, in particular CO_2 [3]. For this purpose, various strategies are used, such as: implementation of cleaner energy sources, reduction of energy demand, reduction of coal trace, capture, storage, or sequestration of CO_2, or using CO_2 as a raw material in further processes. For example, a separated CO_2 can be used in enhanced oil recovery operations (EOR). The next application is the cultivation of algae, phytoplankton, or bacteria, which in turn allows the production of food, nutritional supplements, and livestock feed, as well as lipids, biofuels, methane and even plastic produced by algae, called polyethylene furandicarboxylane (PEF) [4]. However, the most promising are technologies admired for the capture and storage of CO_2. CO_2 storage methods include geological method, deep ocean storage and based on fixation in inorganic carbonates [5].

In turn, the CO_2 capture technology consists in its removal from technological streams or exhaust gases and then its subsequent processing. There are three leading technologies intended for capturing CO_2: pre-combustion, intra-combustion, and post-combustion [6]. The last of these technologies, consisting in separating CO_2 from exhaust gases, can be adapted to existing industrial installations, and the released CO_2 can become a substrate in subsequent production processes.

In recent years, many researchers have focused on a new type of hybrid materials, based on polymer composites with dielectric and/or magnetic fillers. Magneto-responsive colloidal systems (ferrofluids and magnetorheological liquids) and magnetic particles sensitive to the magnetic field have been used in many areas, such as medicine, the processes of separation of chemical, environmental and biological samples [7]. The introduction of magnetic particles into polymer matrices allowed to obtain materials showing new interesting properties [8–10].

The second element that plays a significant role in preparing the appropriate gas separation membranes is to choose the right polymer as a hybrid membrane matrix. It depends on many features, such as: gas permeability and mechanical properties, their cost, etc. Usually, as membrane polymer matrices for CO_2 separation are used various types of polymers, such as polyimides (PI), fluorinated polyimides (FPI), polyamides, polysulfone (PSF), cellulose acetate (CA), polydimethylsiloxane (PDMS), poly(vinyl acetate) (PVAc), poly (2,6-dimethyl-1, 4-phenylene oxide) (PPO), brominated sulfonated poly (2,6-dimethyl-1,4-phenylene oxide (BSPPO), polyvinyl amine, etc. [7, 8]. However, unfortunately, highly selective polymers are not characterized by high permeability and vice versa [11].

Considering the above-mentioned information authors have decided to apply 2 types of hybrid membranes based on modified PPO and PEEK and functionalized carbon nanotubes as inorganic fillers.

Polyphenylene oxide (PPO) finds numerous applications, including in the electrical, electronics and automotive industries. Poly (2,6-dimethyl-1,4-phenylene oxide) (PPO) is also polymer with one of the highest gas permeabilities among aromatic polymers (the chain packing is suppressed (relatively large FFV) by the presence of ether bonds and the absence of polar groups). It was chosen as a matrix due to its excellent mechanical properties, plasticity, and high glass transition temperature. PPO membranes unfortunately show moderate selectivity, caused by the difficult free turnover of the phenyl ring by methyl groups attached on both sides. PPO is also not easily soluble in conventional aprotic solvents, because it is a hydrophobic polymer. In order to improve gas transport by PPO, many electrophilic substitutions were used, such as, for example, bromination, carboxylation, sulfonylation, acylation, etc. [12].

One way to improve the selectivity of the polymer is the introduction of polar groups, inducing stronger interactions between polymer chains, as in research conducted for Nafion, sulfonated polystyrene, polyetherosulfone and poly (ether ether ketone) [13].

PPO sulfonation causes a linear increase in density with ion exchange capacity (IEC), and also improves selectivity, but unfortunately it can reduce gas permeability [14], which could be enhanced by introduction of appropriate filler particles.

As the next polymer matrix, the authors have used PEEK due to its high selectivity in gas separation and excellent thermal, chemical, and mechanical stability [15].

In order to increase the compatibility of an inorganic additive with a polymer matrix and reduce the size of the introduced particles as a new filler, multi-wall nanotubes Fe@MWCNTs were proposed as new carbon fillers. CNTs, due to their unique mechanical, optical, thermal, electrical, and magnetic properties, have found many potential applications in various fields, such as: catalysis, electronics, military technology, energy, material engineering or nanomedicine [10, 29].

Currently, a particularly interesting group are CNTs obtained by the synergy of hybrid material components, i.e., carbon nanostructures and magnetic nanoparticles (CEMNPs and MNPs) [16]. Many carbon-based fillers, such as carbon black (CB), graphene, fullerenes were used as fillers in hybrid membranes, applied for the separation of gas mixtures. Due to the fact that CNTs compared to other carbon allotropes have excellent mechanical properties, smooth surface, and a large specific surface, they will constitute appropriate fillers of inorganic–organic hybrid membranes (better mechanical resistance, numerous selective gas adsorption sites and highly efficient transport gases) [9, 17]. However, MWCNTs tend to aggregate in a polymer matrix (due to the Van der Waals forces and π–π interactions), which can lead to a reduction in gas permeability. This may be due to the fact that nanotubes randomly dispersed in a polymer matrix will cause transport through discontinuous and tortuous paths of interfacial void spaces between MWCNT aggregates and polymer [9, 18, 19]. To maximize the permeability and selectivity of hybrid membranes, it is desirable to disperse MWCNTs homogeneously in the matrix and additionally align them vertically to the membrane surface (Fig. 6.1a). For this purpose, several methods were used, such as the functionalization/modification of CNT external walls (covalent and non-covalent functionalization) and/or polymer matrix, as well as the introduction of an electric field [15, 17, 20]. Due to the fact that Fe@MWCNT tends to aggregate in a polymer matrix, which can lead to reduction of gas transport by membranes, the authors proposed the use of a magnetic field and the functionalization of both inorganic additive and polymer matrix.

The authors have successfully synthesized CO_2 selective novel hybrid membranes, using PPO, SPPO [21] and SPEEK polymer matrices and of course Fe@MWCNTs and modified Fe@MWCNT-OH with increased iron content (5.80 wt%.) [22]. It was found that the introduction of such nanofiller particles positively influenced gas transport properties through increase of CO_2 diffusion coefficient D, permeation coefficient P, sorption S and selectivity α_{CO_2/N_2}. Also, magnetic, thermal, and mechanical parameters of the analyzed membranes were enhanced, especially after the use of magnetic casting and chemical modification of the inorganic and organic phases. The application of the magnetic field in the production of hybrid membranes enhanced the proper arrangement and improvement of CNT dispersion in the polymer matrix and had a positive effect on their gas transport properties. Also, the introduction of polymers with paramagnetic and CNTs with slightly ferromagnetic properties for the production of hybrid membranes made it possible to obtain their positive response to the applied magnetic field (Fig. 6.1c).

In turn, the chemical modification of both polymer matrices and carbon nanotubes allowed to increase the compatibility between these phases and significantly increased the CO_2 separation efficiency (Fig. 6.1e), through rise of the selectivity coefficient α_{CO_2/N_2} (44.28 and 58.63 for Fe@MWCNT-OH/SPPO and Fe@MWCNT-OH/SPEEK, respectively) and the permeation coefficient P_{CO_2} values (86.02 and 59.56. Barrer for Fe@MWCNT-OH/SPPO and Fe@MWCNT-OH/SPEEK, respectively). Thus, a positive impact of both the chemical modification

6.1 New Directions of Research in the Field of Clean Coal Technologies

of the polymer matrices and inorganic additives on the CO_2 transport properties were clearly found.

It has been assumed that the presence of polar sulfonic and sulfonate groups can promote CO_2-sulfone interactions, enabling favorable affinity to CO_2. While better phase compatibility may reduce the defects at their boundary and regulate the appropriate fractional free volume (FFV) to increase the CO_2 transport. On the other hand, the introduction of modified CNTs, forming "supra-structures" in the polymer matrix, can increase the rigidity of the polymer chains, extend the CO_2-philic pathways and increase the interaction with CO_2 electric quadrupole moment.

It was also stated (Fig. 6.1d) that the increase of modified CNTs addition caused the improvement of membrane's mechanical parameters [increase in R_m from 33.09 to 68.64 MPa for hybrid SPEEK membranes and from 25.83 to 61.89 MPa for hybrid SPPO and in E from 1.17 to 2.08 GPa (hybrid SPEEK) and from 1.07 to 2.21 GPa (hybrid SPPO)] and their thermo-oxidative stability (Fig. 6.1b).

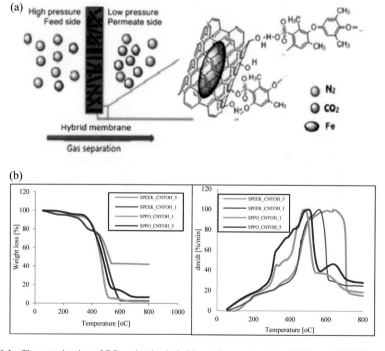

Fig. 6.1 Characterization of CO_2 selective hybrid membranes based on SPPO and SPEEK matrices and functionalized carbon nanotubes Fe@MWCNT-OH: **a** scheme of hybrid membranes preparation and gas permeation process, **b** TGA and DTG results for hybrid membranes with various Fe@MWCNT-OH addition, **c** magnetic properties of hybrid membranes, hysteresis loops of polymer and hybrid membranes with various Fe@MWCNT-OH additions, **d** mechanical properties of hybrid membranes: R_m and E versus Fe@MWCNT-OH loadings for hybrid SPEEK and SPPO membranes cast in the presence of stronger magnetic field, **e** selectivity coefficient α_{CO_2/N_2} versus permeation coefficient P_{CO_2} regarding the Robeson upper bound line

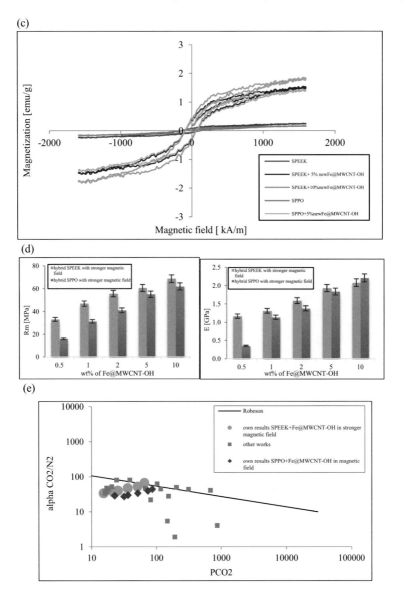

Fig. 6.1 (continued)

Based on the analysis of experimental data using a computer application created by the authors, it was found that the most suitable model for describing CO_2 transport through the analyzed hybrid membranes was the model developed by Chehrazi et al. (% AARE error was 12–16%). It was stated the reduction of interphase thickness and the increase of ratio P_{NT}/P_m with increasing magnetic field induction, which indicates better interfacial interaction, compatibility, and gas transport properties of analyzed membranes.

The authors have stated that when comparing the parameters of the proposed membranes with other hybrid membranes (Table 6.1), they are characterized by significantly improved properties or are comparable to the membranes representing the latest trends in CO_2 separation. It means that these materials can be successfully used in the power industry in the future.

In addition to generating greenhouse gases (especially CO_2), during the combustion of fossil fuels, large amounts of solid waste in the form of fly ashes are produced (over 750 million tons annually), of which only 30% is used [33]. However, when analyzing their composition, it can be stated that they can become an excellent secondary source of rare earth elements (REEs). This is particularly important, considering their extraordinary properties (chemical, catalytic, physical, magnetic, and luminescent), limited availability and numerous applications, especially in modern technologies [34, 35]. Various methods are used to recover REEs from ash, such as chemical, biological, and physical [36]. The most effective method is chemical recovery, most often consisting of several stages, such as acid–base leaching and then separation of individual REEs from the obtained extracts. However, these

Table 6.1 Comparison of permeation and selectivity coefficients of hybrid membranes based on several types of CNTs and polymer matrices

Hybrid membrane polymer matrix	Filler of hybrid membrane	CNT [wt%]	Pressure [bar]	Temperature [°C]	P_{CO_2} [Barrer]	alpha CO_2/N_2	Refs.
BTDA-TDI/MDI (P84)	MWCNT	2	1	25	190.5	1.90	[23]
Polyimide	CNT	15	1	25	866.6	4.10	[24]
Cellulose acetate	CNT	0.1	3	-	147	5.50	[25]
6FDA-TP polyimide	AP-SWNT	2	16.4	35	81	22.00	[26]
PSF	CNT/ZIF-301(30)	2	2	25	16	34.00	[27]
Polyimide	MWCNT-COOH-OH	3	1	15	9.06	37.70	[28]
Pebax	MWCNT	15	7	34.85	680	41.00	[29]
CMC	CNT	1	2	80	116	45.00	[30]
TFN	CNT-GO	1:1	4	70	66.3	47.10	[31]
PC/PEG	C-MWCNT	10	2	25	20.32	52.10	[32]
SPPO	Fe@MWCNT-OH	10	5	25	86.02	44.28	Own work [21]
SPEEK	Fe@MWCNT-OH	10	5	25	59.56	58.63	Own work [22]

extracts have complex composition, extreme pH, and high content of matrix components (Si, Fe, Al) [37]. Conventional methods (ion exchange, coagulation, flocculation, flotation, adsorption, and chemical precipitation) are used for this purpose, but they are characterized by high energy consumption, low selectivity, regeneration problems, high environmental and operational costs [38–40]. Therefore, alternative techniques are sought, based on chelating agents, ionic liquids, and membranes [41–45]. Researchers have used various membrane techniques for the recovery of REEs from extracts, such as nanofiltration (NF), reverse osmosis (RO), emulsification liquid membranes (ELM) and hollow fiber liquid membranes (HFLM). Most of the proposed membrane techniques allowed to obtain REE enrichment above 90%, especially in the case of NF-based techniques, as well as ELM and HFLM (almost 100%) [43–50]. However, only some membrane techniques, such as multi-stage HFLM or based on ion-imprinted polymer membranes, allow the recovery of single REE, in contrast to the rest, in which REEs concentrates were in the form of the ion mixtures [49–51].

In the case of membrane technology, selective recovery of metal ions from these complex mixtures can be easily achieved only using modern materials with high selectivity and REE affinity, such as, for example, ion-imprinted polymers (IIPs), which have specific sites in the polymer matrix, designed for binding appropriate ions. IIPs are easy to synthesize, stable, selective in the presence of other ions, inexpensive and show the possibility of repeated use. Until now, ion-imprinted polymers were used as a solid phase in extraction technique SPE for preconcentration of Eu, Ce, Nd, Dy, Y ions from sewage samples, biological samples, like human serum and plasma or as sensors [52, 53]. They are cross-linked polymers with characteristic binding sites for target ions, which can be synthesized in several stages during the reaction of the functional monomer, crosslinker, initiator and template ion [54, 55]. In the first stage, complexes based on monomers with functional groups and REE ions are created. In the second stage, photo- or thermal polymerization of monomers takes place. And finally in the last stage, the template REE ions are leaked from the polymer matrix, which leads to the formation of specific binding sites, which can later capture REE ions from the solution [56]. Of course, their adsorption ability depends on many factors, such as the ability of ligands to bind metal ions, ions size, their charge, metal electron configuration and the degree of oxidation. These types of materials are characterized by adequate pH and thermal stability. IIPs operate on the principle of a key and lock mechanism to recognize and remove target ions, even in trace amounts, which was not possible using other methods [57, 58]. These polymers can be synthesized by creating binary or ternary complexes. However, current studies are heading towards IIPs synthesis, based on ternary complexes, as during the copolymerization of the complex between REE (III), 5,7-dichloroquinoline-8-ol and 4-vinylpyridine (4-VP) and styrene-DVB and also other systems based on HEMA-EDMA and MMA-EDMA as monomers and crosslinkers [59–65]. The next type of IIPs with strong REE affinity have been synthesized using REE ions and complexing agent, namely Schiff base ligand in the presence of EGDMA (cross-linking agent), 4-VP (functional monomer), and azobis (isobutyronitryl) (AIBN) as initiator [66,

6.1 New Directions of Research in the Field of Clean Coal Technologies

67]. It was sated that the maximum REE recovery that was obtained did not exceed 80% (DCQ leakage problem) [68].

Magnetic adsorbents in the form of Fe_3O_4, $Fe_3O_4@SiO_2$, CNT and MWCNT nanoparticles [69] are also used in the SPE extraction. However, currently rather multi-walled carbon nanotubes (MWCNTs), which have a large surface, high mechanical strength and chemical stability are increasingly used as a solid phase for analysis of Pb, Au, Rh, Mn, Fe, Cu and REEs in environmental samples, like mineral water and synthetic sea water, biological samples, such as serum or garlic and geological samples, such as rocks. Despite their excellent properties, but due to their stiffness, chemical unreactivity and the ability to aggregate, CNTs are difficult to dissolve or disperse in organic solvents or polymer matrices, which seriously limits their potential use in various techniques. A large effort is currently focused on the use of covalent or non-covalent functionalization methods to improve their solubility, such as modification with Congo red (CR) or polyacrylic acid (PAA) (Eu (III) adsorption) [70, 71]. In addition, oxidation increases CNT's adsorption capacity, which is a key factor for metal sorbents. Our recent studies have also confirmed the positive impact of the external magnetic field application on improvement of Fe@MWCNTs dispersion in a polymer matrix [72]. However, in most cases, magnetic molecules were used for adsorption of previously formed REE complexes, such as LA-PAR adsorbed on a carbon-based magnetic nanocomposite (AC-MNC) [73].

In order to examine the problem of recovery of selected REE ions, namely Nd, Y and Gd from synthetic extracts of coal fly ashes, the authors proposed two types of membranes. Namely, the first type was based on modified chitosan, imprinted with Nd or Y ions, designed to separate Nd or Y from solutions corresponding to the composition of real extracts [74]. The second type were Gd selective, hybrid membranes with a complex matrix, consisting of a modified chitosan imprinted with Gd ions with addition of GdIIPs and filler in the form of carbon nanotubes modified with CuNiCo ferrite [75].

In the first stage, a modified chitosan (Schiff base) and ion imprinted copolymer based on the Gd (III) complex, 5,7-dichloroquinoline-8-ol and 4-vinylpyridine as well as styrene-DVB were obtained (Fig. 6.2a). Next chitosan solution was prepared, to which neodymium (III) nitrate, yttrium (III) nitrate or gadolinium (III) nitrate were added. In the case of hybrid membranes was used complex matrix was used consisting of modified chitosan and 30% of Gd imprinted polymer with addition of 0.5–5% modified MWCNTs. After adding a crosslinker, the polymer or hybrid membranes were poured on levelled Petri dishes in the external magnetic field and the solvent was evaporated. Then the membrane was leached with HCl solution to remove REE ions and obtain specific binding sites. The last step was the immersion of the membrane in sodium hydroxide solution.

The created membranes were studied in terms of their use in the recovery of selected REE ions for this purpose by conducting adsorption tests, taking into account the impact of pH, analysis of kinetics and isotherms. Selectivity and reuse of synthesized materials were also tested. Additionally, the analysis of separation capacity for Gd imprinted membranes was performed using Sterlitech HP4750 high pressure stirred cell kit.

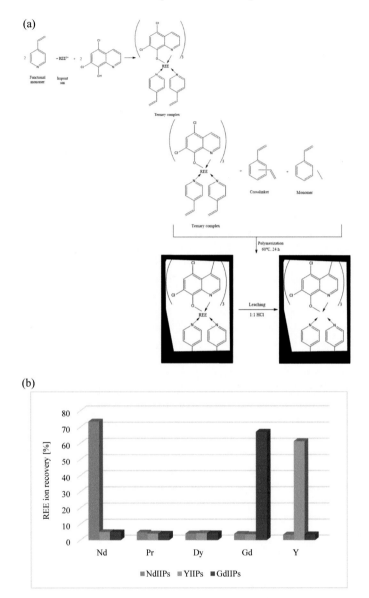

Fig. 6.2 Characterization of REE selective membranes based on ion imprinted polymers and hybrid membranes with modified MWCNTs addition: **a** scheme of multi-stage synthesis of REE ion imprinted polymer, **b** the recovery of REE ions from adsorptive membranes after the separation process, **c** dependence of retention coefficient of Gd selective hybrid membranes versus amount of MWCNT addition, **d** magnetic properties of REE selective membranes, hysteresis loops of polymer and hybrid membranes

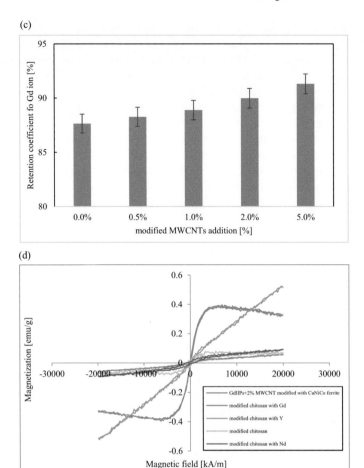

Fig. 6.2 (continued)

The generated experimental data were analyzed using two computer applications, namely REE 2.0 (based on Lagergren pseudo-first and pseudo-second order, Elovich kinetic models, intraparticle diffusion, diffusion-chemisorption and Boyd models) and REE_isotherm (based on Langmuir, Freundlich and Dubinin-Radushkevich models) created earlier by authors [76, 77]. It was stated that the adsorption of the analyzed Nd, Y and Gd ions in the obtained membranes occurred in accordance with the mechanism of chemisorption and was mainly controlled by film diffusion. The specific binding sites on the adsorbent surface were evenly distributed, and the adsorption membranes showed a strong affinity for the tested ions. It was also found that the adsorption of Nd and Y ions showed the characteristics of regular monolayer adsorption.

The selectivity tests carried out for the membranes based on a modified chitosan imprinted with Nd and Y ions showed that the distribution coefficient K_D was 1472.8 and 964.64 ml/g, respectively. However, even higher value of this distribution coefficient K_D was obtained for Gd imprinted hybrid membranes with the addition of functionalized MWCNTs (2042.81 ml/g). Such high values of considered parameter translate into high efficiency of the proposed membranes in REE ions recovery, especially in the case of hybrid membranes.

In turn, during the multiple use of membranes, a slight decrease in adsorption capacity after the first cycle was recorded, but still over 85% of the value of 1 cycle after another 5 cycles was obtained. The thermal stability of the analyzed materials after modification was also improved.

The produced membranes showed a high level of rejection of the matrix components (Na, Mg, Ca, Al, Fe and Si) in synthetic extracts of coal fly ashes, while the retention ratio of Nd and Y ions was 90.11% and 80.95%, respectively. While the retention coefficients of hybrid membranes imprinted with Gd ion increased with the CNTs addition, finally reaching 91.30% for the largest inorganic addition (Fig. 6.2c).

In addition, the recovery of REE ions using the analyzed membranes for Nd and Y ions was from 61.38 to 73.50% (Fig. 6.2b). This may be due to the presence of the remaining four REE ions in the extract, which, due to their chemical similarity, are partially adsorbed on the adsorption membranes and constitute a certain percentage of the recovered solution. This can be solved by further modification of the membranes, as in the case of hybrid membranes, where the recovery of Gd ions was much higher and amounted to 82.70%. All this indicates the possibility of potential application of synthesized IIPs as adsorption membranes for the recovery of Nd, Y and Gd ions, in particular hybrid membranes based on 2 different Gd imprinted polymers with the addition of modified CNTs.

Also due to the use of an external magnetic field in the preparation of hybrid membranes, their magnetic properties were examined (Fig. 6.2d). It was found that the shape of the hysteresis loops of examined membranes based on modified chitosan imprinted with Nd and Gd ions were typical for weak ferromagnetic materials. However, in the case of chitosan imprinted with yttrium ions, higher magnetization values were discovered, and the shape of the loop indicated a transition towards superparamagnetic materials. While the introduction of carbon nanotubes modified with CuNiCo ferrite significantly increased membrane's magnetization. And this in turn had a direct impact on the better dispersion of the inorganic additive in the polymer matrix and adsorption and separation properties of the examined membranes.

It was also performed the comparison of adsorption capacity of synthesized new adsorptive membranes with other IIPs membranes (Table 6.2) and it was found that they were characterized by higher or comparable adsorption capacity. Of course, it must be considered that this parameter depends on the form of the tested materials and conditions of adsorption process (pH, temperature). Except that the proposed membranes are produced of biodegradable chitosan, which indicates their harmlessness and ecological potential. That is why the authors have continued the work on them and introduced the hybrid membranes based on proposed polymer matrices and functionalized carbon nanotubes, obtaining membranes with even better adsorption

6.1 New Directions of Research in the Field of Clean Coal Technologies

properties. All this indicates the huge potential of the proposed membrane materials and the possibility of their further development and the possibility of potential implementation in industry.

Table 6.2 Comparison of adsorption capacity of proposed adsorptive hybrid membranes with other REE ion imprinted materials

Ion imprinted material	Cation	Adsorption capacity [mg/g]	References
Molecularly imprinted polymers	Nd (III)	14.60	[78]
IMCFs	Nd (III)	22.61	[79]
Ionic imprinted polymer particles	Nd (III)	33.00	[80]
Dual template docking oriented ionic imprinted mesoporous films	Nd (III)	34.98	[81]
Phosphonic acid functionalized silica microspheres	Nd (III)	45.00	[82]
Magnetic chitosan nano-based particles grafted with amino acids	Dy (III)	8.90–17.60	[83]
Ion imprinted mesoporous silica materials	Dy (III)	22.30	[84]
II-MAC	Dy (III)	23.30	[53]
Imprinted styrene–divinylbenzene copolymer	Dy (III)	40.15	[63]
Diethylenetriamine-functionalised chitosan magnetic nano based particles	Dy (III)	51.50	[85]
Imprinted polymer using Schiff base	Ce (III)	24.50	[68]
Imprinted polymer using 4-(2,4-dihydroxyphenazylo) acetophenone	Ce (III)	24.70	[86]
Imprinted cryogel using N-methacryloamido antipyrine as functional monomer	Ce (III)	36.58	[87]
Sc (III) ion imprinted polymers	Sc (III)	12.80	[88]
IIP-HQP/SiO$_2$	Pr (III)	18.32	[89]
Ionic imprinted resins based on EDTA and DTPA derivatives	Gd (III)	24.53	[90]
NdIIPs based on modified chitosan	Nd (III)	40.80	Own work [74]
YIIPs based on modified chitosan	Y(III)	36.20	Own work [74]
GdIIPs based on modified chitosan with modified CNTs addition	Gd (III)	41.81	Own work [75]

6.2 Challenges and Opportunities for Clean Coal Technologies

The SWOT analysis for clean coal technologies conducted by the authors allowed for the identification of the strengths, weaknesses of CCTs, as well as opportunities and threats to their development located in the macroenvironment of energy systems.

The most important strengths of CCT include technological advances made in the development and improvement of existing solutions. This has increased the efficiency of many of them, such as ultra-supercritical power plants, gasification methods, integrated gas-steam cycles, or membrane techniques. The importance of coal for stable energy supplies is also important, e.g. in China, India, the USA, or Poland. The weaknesses of CCT include the concentration of leading technologies on reducing CO_2 emissions. They do not take into account other harmful gases such as NO_x, SO_x, or heavy metals. Furthermore, CCS allows for the reduction of carbon dioxide emissions, but it is usually not 100%. Some CCTs are also energy-intensive, which can lead to a decrease in the energy efficiency of power plants. CCS requires access to the appropriate geological formations that allow CO_2 storage. However, it should be noted that each of the weaknesses of CCT can be eliminated by combining several of them, e.g. supercritical combustion and CCS, CCUS, the use of membrane techniques, and building energy clusters. The costs of the technology may also be an obstacle to the development of CCTs, but, similarly to RES, they will certainly be reduced in the case of mass implementation of CCT solutions.

On the side of opportunities, the SWOT analysis identified a growing demand for energy, which in many cases can be generated solely on the basis of coal, as in the case of China, where coal-fired power plants are the basis of the energy system and this state will be maintained in the near future. In connection with the adopted environmental standards, it is necessary to apply solutions that allow for reaching a compromise in this aspect. China is actively working on the development of CCS, and in the future it may become a supplier of this technology, as is the case with RES. The aspect of the possibility of building hybrid systems, combining CCT and RES, is also important. Technological progress and the support of governments such as China and the United States are important opportunities. Recent changes in the narrative of the EU on CCTs may also constitute an opportunity for their development. On the side of threats, competition from other energy sources was identified, mainly RES. Renewable energy is perceived as a substitute for energy obtained on the basis of fossil fuels, which is supported by falling prices of RES technologies. Another important factor is climate policy, which will decide about the future of CCT in the coming years. Another threat may be the negative attitude of society towards CCTs, which rely on fossil fuels. These are often seen as a source of threat to the natural environment.

6.3 The Future of Clean Coal Technologies

The future of clean coal technologies will depend on the factors identified during the SWOT analysis. If the weaknesses of technology and threats from the macroenvironment are eliminated, CCTs have a chance for further development and industrial-scale existence in the energy systems of the stakeholder countries. Hybrid energy systems are a particularly promising solution. By combining CCTs with renewable energy sources, it is possible to achieve synergy effect. Together, they have the ability to form a stable energy system, enable the flexibility, and facilitate the management of energy generated by fossil fuels and renewable energy sources. This will also facilitate the management of not only energy, but also other resources obtained from coal and RES, such as CO_2, NO_x, SO_2, REEs, biomass, hydrogen, syngas, and liquid fuels. Such a solution has huge potential in the field of energy transition, achieving climate neutrality, and ensuring energy security.

To make this possible, it is necessary to use modern solutions, for example, in the form of membrane techniques based on unusual methods and new membrane materials. During their work, the authors proposed innovative solutions regarding both carbon dioxide separation from the obtained gas mixture, as well as the processing of solid waste in the form of coal fly ashes in the proposed CCT technologies. Hybrid membranes intended for CO_2 separation were based on modified polymer matrices and functionalized carbon nanotubes Fe@MWCNTs, which made it possible to obtain materials with improved separation properties as well as thermal and mechanical features. The introduced modifications caused an increase in the selectivity coefficient by up to 70% and CO_2 permeability coefficient almost four times.

In turn, membranes intended for the recovery of selected rare earth metal ions from the extracts of coal fly ashes are based on modern selective ion-imprinted polymers and functionalized carbon nanotubes modified with CuNiCo ferrite. The introduced modifications enabled better dispersion and interaction between organic and inorganic phase of a hybrid membrane. In this way, materials with improved separation properties were obtained, where the recovery of selected REE ions reached more than 82% and retention ratio was even over 90%, as well as much better mechanical and thermal stability. Also, the use of a polymer matrix and carbon nanotubes with magnetic properties, as well as the introduction of an external magnetic field during the production of the used hybrid membranes, significantly improved their properties. The significant improvement of hybrid membrane transport and separation properties may result from the synergistic effect of modified MWCNTs and polymer matrices and indicate their future potential application.

So it can be seen that the future lies in the introduction of advanced materials, such as modern polymer materials and nanomaterials CNTs and technologies using artificial intelligence AI and machine learning, allowing optimization and improvement of the proposed systems.

It is certain that the abrupt removal of coal (or other fossil fuels) from the mixes of countries that still rely on it would lead to multidimensional effects. First, to disruptions in industrial production, construction, and steel production, which would have a negative impact on the economy. Not only in coal regions, the recession lead to mass layoffs, unemployment, and social protests. The resulting energy shortages would destabilise the functioning of citizens, businesses, and state services. They would also destabilise energy, social, and military security. Such dramatic changes would affect greenhouse gas emissions. Initially, they would certainly decrease, but the need to search for coal substitutes would encourage society to use, for example, other fossil fuels, which would delay the achievement of climate goals. Therefore, since removing coal overnight is impossible and will have to be a gradual process, the use of CCT seems to be the only sensible solution that will allow for a smooth and fully controlled transition to zero-emission energy.

The field of clean coal technologies is recently developing rapidly, and new research focuses on improving carbon dioxide capture, improving gasification and combustion processes, optimization of waste management and integration with renewable energy sources. These innovations have the potential to make hard coal a more balanced energy source, while dealing with environmental and economic challenges. Further research and development, combined with supporting investments, will be crucial for the development of these technologies and achieving a cleaner energy future.

References

1. Fang S, Rahaman M, Bharti J, Reisner E, Robert M, Ozin GA, Hu YH (2023) Photocatalytic CO_2 reduction. Nat Rev Methods Primers 3(1):61
2. Sattari A, Ramazani A, Aghahosseini H, Aroua MK (2021) The application of polymer containing materials in CO_2 capturing via absorption and adsorption methods. J CO_2 Util 48:101526
3. Han Y, Ho WSW (2021) Polymeric membranes for CO_2 separation and capture. J Membr Sci 628:119–244
4. Kokkoli A, Zhan Y, Angelidaki I (2018) Microbial electrochemical separation of CO_2 for biogas upgrading. Biores Technol 247:380–386
5. Huang K, Zhang J-Y, Liu F, Dai S (2018) Synthesis of porous polymeric catalysts for the conversion of carbon dioxide. ACS Catal 8(10):9079–9102
6. Bryan N, Lasseuguette E, van Dalen M, Permogorov N, Amieiro A, Brandani S, Ferrari M-C (2014) Development of mixed matrix membranes containing zeolites for post-combustion carbon capture. Energy Procedia 63:160–166
7. Huang J, Li X, Luo L, Wang H, Wang X, Li K (2014) Releasing silica-confined macromolecular crystallization to enhance mechanical properties of polyimide/silica hybrid fibers. Compos Sci Technol 101:24–31
8. Chandrappa KG, Venkatesha TV (2012) Generation of Co_3O_4 microparticles by solution combustion method and its Zn–Co_3O_4 composite thin films for corrosion protection. J Alloy Compd 542:68–77
9. Boncel S, Herman AP, Walczak KZ (2012) Magnetic carbon nanostructures in medicine. J Mater Chem 22:31–37
10. Bok-Badura J, Jakobik-Kolon A, Turek M, Boncel S, Karon K (2015) A versatile method for direct determination of iron content in multi-wall carbon nanotubes by inductively coupled plasma atomic emission spectrometry with slurry sample introduction. RSC Adv 5:101634–101640

References

11. Robeson LM, Freeman BD, Paul DR, Rowe BW (2009) An empirical correlation of gas permeability and permselectivity in polymers and its theoretical basis. J Membr Sci 341:178–185
12. Chowdhury G, Kruczek B, Matsuura T (2001) Polyphenylene oxide and modified polyphenylene oxide membranes. In: Gas, vapor and liquid separation. Springer Science + Business Media, pp 199–214
13. Chen C, Martin C (1994) Gas-transport properties of sulfonated polystyrenes. J Membr Sci 9:51–61
14. Kruczek B, Matsuura T (2000) Effect of metal substitution of high molecular weight sulfonated polyphenylene oxide membranes on their gas separation performance. J Membr Sci 167:203–216
15. Wu H, Shen X, Xu T, Hou W, Jiang Z (2012) Sulfonated poly (ether ether ketone)/amino-acid functionalized titania hybrid proton conductive membranes. J Power Sour 213:83–92
16. Liu S, Boeshore S, Fernandez A, Sayagues MJ, Fischer JE, Gedanken A (2001) Study of cobalt-filled carbon nanoflasks. J Phys Chem B 105(34):7606–7611
17. Oueiny C, Berlioz S, Perrin F-X (2014) Carbon nanotube–polyaniline composites. Prog Polym Sci 39(4):707–748
18. Wu B, Li X, An D, Zhao S, Wang Y (2014) Electro-casting aligned MWCNTs/polystyrene composite membranes for enhanced gas separation performance. J Membr Sci 462:62–68
19. Bystrzejewski M, Huczko A, Lange H (2005) Arc plasma route to carbon-encapsulated magnetic nanoparticles for biomedical applications. Sens Actuat B Chem 109(1):81–86
20. Al-Mufti SMS, Ali MM, Rizvi SJA (2020) Synthesis and structural properties of sulfonated poly ether ether ketone (SPEEK) and poly ether ether ketone (PEEK). AIP Conf Proc 2220:020009
21. Rybak A, Rybak A, Kaszuwara W, Boncel S, Kolanowska A (2022) Characteristics of inorganic-organic hybrid membranes containing carbon nanotubes with increased iron-encapsulated content for CO_2 separation. Membranes 12(2):13367–13380
22. Rybak A, Rybak A, Boncel S, Kolanowska A, Kaszuwara W, Kolev SD (2022) Hybrid organic-inorganic membranes based on sulfonated poly (ether ether ketone) matrix and iron-encapsulated carbon nanotubes and their application in CO_2 separation. RSC Adv 12(21):13367–13380
23. Sazali N, Salleh WNW, Ismail AF, Wong KC, Iwamoto Y (2018) Exploiting pyrolysis protocols on BTDA-TDI/MDI (P84) polyimide/nanocrystalline cellulose carbon membrane for gas separations. J Appl Polym Sci 136:136
24. Tseng H-H, Kumar IA, Weng T-H, Lu C-Y, Wey M-Y (2009) Preparation, and characterization of carbon molecular sieve membranes for gas separation—the effect of incorporated multi-wall carbon nanotubes. Desalination 240(1–3):40–45
25. Ahmad A, Jawad Z, Low S, Zein S (2014) A cellulose acetate/multi-walled carbon nanotube mixed matrix membrane for CO_2/N_2 separation. J Membr Sci 451:55–66
26. Zhang Q, Li S, Wang C, Chang H-C, Guo R (2020) Carbon nanotube-based mixed-matrix membranes with supramolecularly engineered interface for enhanced gas separation performance. J Membr Sci 598:117794
27. Sarfraz M, Ba-Shammakh M (2018) Harmonious interaction of incorporating CNTs and zeolitic imidazole frameworks into polysulfone to prepare high performance MMMs for CO_2 separation from humidified post-combustion gases. Braz J Chem Eng 35:217–228
28. Sun H, Wang T, Xu Y, Gao W, Li P, Niu QJ (2017) Fabrication of polyimide and functionalized multi-walled carbon nanotubes mixed matrix membranes by in situ polymerization for CO_2 separation. Sep Purif Technol 177:327–336
29. Zhao D, Ren J, Wang Y, Qiu Y, Li H, Hua K, Li X, Ji J, Deng M (2017) High CO_2 separation performance of Pebax®/CNTs/GTA mixed matrix membranes. J Membr Sci 521:104–113
30. Borgohain R, Jain N, Prasad B, Mandal B, Su B (2019) Carboxymethyl chitosan/carbon nanotubes mixed matrix membranes for CO_2 separation. React Funct Polym 143:104331
31. Wong KC, Goh PS, Taniguchi T, Ismail AF, Zahri K (2019) The role of geometrically different carbon-based fillers on the formation and gas separation performance of nanocomposite membranes. Carbon 149:33–44

32. Moghadassi AR, Rajabi Z, Hosseini SM, Mohammadi M (2013) Preparation and characterization of polycarbonate-blend-raw/functionalized multi-walled carbon nanotubes mixed matrix membrane for CO_2 separation. Sep Sci Technol 48:1261–1271
33. Baba A, Usmen MA (2006) Effects of fly ash from coal-burning electrical utilities on ecosystem and utilization of fly ash. In: Groundwater ecosystems, pp 15–31
34. Balaram V (2019) Rare earth elements: a review of applications, occurrence, exploration, analysis, recycling, and environmental impact. Geosci Front 10:1285–1303
35. Taggart RK, Hower JC, Hsu-Kim H (2018) Effects of roasting additives and leaching parameters on the extraction of rare earth elements from coal fly ash. Int J Coal Geol 196:106–114
36. Muravyov MI, Bulaev AG, Melamud VS, Kondrateva TF (2015) Leaching of rare earth elements from coal ashes using acidophilic chemolithotrophic microbial communities. Microbiology 84:194–201
37. Nascimento M, Moreira Soares PS, de Souza VP (2009) Adsorption of heavy metal cations using coal fly ash modified by hydrothermal method. Fuel 88(9):1714–1719
38. Sahoo PK, Kim K, Powell MA, Equeenuddin SM (2016) Recovery of metals and other beneficial products from coal fly ash: a sustainable approach for fly ash management. Int J Coal Sci Technol 3:267–283
39. Mutlu BK, Cantoni B, Turolla A, Antonelli M, Hsu-Kim H, Wiesner MR (2018) Application of nanofiltration for rare earth elements recovery from coal fly ash leachate: performance and cost evaluation. Chem Eng J 349:309–317
40. Gasser MS, Aly MI (2013) Separation and recovery of rare earth elements from spent nickel-metal-hydride batteries using synthetic adsorbent. Int J Miner Process 121:31–38
41. Huang C, Wang Y, Huang B, Dong Y, Sun X (2019) The recovery of rare earth elements from coal combustion products by ionic liquids. Miner Eng 130:142–147
42. Hasegawa H, Rahman IMM, Egawa Y, Sawai H, Begum ZA, Maki T, Mizutani S (2014) Recovery of the rare metals from various waste ashes with the aid of temperature and ultrasound irradiation using chelants. Water Air Soil Pollut 225:2112
43. Shimizu H, Ikeda K, Kamiyama Y (1992) Refining of a rare earth including a process for separation by a reverse osmosis membrane. U.S. Patent No. 5,104,544
44. Wen B, Shan X, Xu S (1999) Preconcentration of ultratrace rare earth elements in seawater with 8-hydroxyquinoline immobilized polyacrylonitrile hollow fiber membrane for determination by inductively coupled plasma mass spectrometry. Analyst 124:621–626
45. Murthy ZVP, Gaikwad MS (2013) Separation of praseodymium (III) from aqueous solutions by nanofiltration. Can Metall Q 52:18–22
46. Innocenzi V, De Michelis I, Ferella F, Vegliò F (2013) Recovery of yttrium from cathode ray tubes and lamps' fluorescent powders: experimental results and economic simulation. Waste Manage 33:2390–2396
47. Qin QW, Zhao HQ, Lai YQ, Li J, Liu YX, Deng ZG (2002) Extraction of rare earth metals by liquid surfactant membranes containing Cyanex272 as a carrier. Mining Metallurgy Eng 22:74–78
48. Davoodi-Nasab P, Rahbar-Kelishami A, Safdari J, Abolghasemi H (2017) Application of emulsion nanofluids membrane for the extraction of gadolinium using response surface methodology. J Mol Liq 244:368–373
49. Wannachod P, Chaturabul S, Pancharoen U, Lothongkum AW, Patthaveekongka W (2011) The effective recovery of praseodymium from mixed rare earths via a hollow fiber supported liquid membrane and its mass transfer related. J Alloy Compd 509:354–361
50. Ambare DN, Ansari SA, Anitha M, Kandwal P, Singh DK, Singh H, Mohapatra PK (2013) Non-dispersive solvent extraction of neodymium using a hollow fiber contactor: mass transfer and modeling studies. J Membr Sci 446:106–112
51. Rybak A, Rybak A (2021) Characteristics of some selected methods of rare earth elements recovery from coal fly ashes. Metals 11(1):142–169
52. Liu E, Xu X, Zheng X, Zhang F, Liu E, Li C (2017) An ion imprinted macroporous chitosan membrane for efficiently selective adsorption of dysprosium. Sep Purif Technol 189:288–295

53. Hande PE, Samui AB, Kulkarni PS (2015) Highly selective monitoring of metals by using ion-imprinted polymers. Environ Sci Pollut Res 22:7375–7404
54. Fu J, Chen L, Li J, Zhang Z (2015) Current status and challenges of ion imprinting. J Mater Chem A 3:13598–13627
55. Hu Y, Pan J, Zhang K, Lian H, Li G (2013) Novel applications of molecularly imprinted polymers in sample preparation. Trends Anal Chem 43:37–52
56. Branger C, Meouche W, Margaillan A (2013) Recent advances on ion-imprinted polymers. React Funct Polym 73:859–875
57. Lu J, Qin Y, Wu Y, Meng M, Yan Y, Li C (2019) Recent advances in ion-imprinted membranes: separation and detection via ion-selective recognition. Environ Sci Water Res Technol 5(1):1626–1653
58. Kusumkar VV, Galamboš M, Viglašová E, Dano M, Šmelková J (2021) Ion-imprinted polymers: synthesis, characterization, and adsorption of radionuclides. Materials 14:1083–1111
59. Moussa M, Pichon V, Mariet C, Vercouter T, Delaunay N (2016) Potential of ion imprinted polymers synthesized by trapping approach for selective solid phase extraction of lanthanides. Talanta 161:459–468
60. Moussa M, Ndiaye MM, Pinta T, Pichon V, Vercouter T, Delaunay N (2017) Selective solid phase extraction of lanthanides from tap and river waters with ion imprinted polymers. Anal Chim Acta 2017(96):344–352
61. Sarabadani P, Sadeghi M, Payehghadr M, Eshaghic Z (2019) Synthesis and characterization of a novel nanostructured ion-imprinted polymer for pre-concentration of Y(III). Mater Res Express 6(11):115306
62. Biju VM, Gladis JM, Rao TP (2003) Ion imprinted polymer particles: synthesis, characterization and dysprosium ion uptake properties suitable for analytical applications. Anal Chim Acta 478:43–51
63. Ramakrishnan K, Rao TP (2006) Ion imprinted polymer solid phase extraction (IIP-SPE) for preconcentrative separation of erbium (III) from adjacent lanthanides and yttrium. Sep Sci Technol 41:233–246
64. Kala R, Gladis JM, Rao TP (2004) Preconcentrative separation of erbium from Y, Dy, Ho, Tb and Tm by using ion imprinted polymer particles via solid phase extraction. Anal Chim Acta 518:143–150
65. Lai X, Hu Y, Fu Y, Wang L, Xiong J (2012) Synthesis and characterization of Lu (III) ion imprinted polymer. J Inorg Organomet Polym 22:112–118
66. Rahman ML, Puah PY, Sarjadi MS, Arshad SE, Musta B, Sarkar SM (2019) Ion-imprinted polymer for selective separation of cerium (III) ions from rare earth mixture. J Nanosci Nanotechnol 19:5796–5802
67. Rohani N, Mustapa N, Fazli N, Malek A, Yusoff MM, Rahman ML (2016) Ion imprinted polymers for selective recognition and separation of lanthanum and cerium ions from other lanthanides. Sep Sci Technol 51(17):2762–2771
68. Giakisikli G, Anthemidis AN (2013) Magnetic materials as sorbents for metal/metalloid preconcentration and/or separation: a review. Anal Chim Acta 789:1–16
69. Su S, Chen B, He N, Hu B, Xiao Z (2014) Determination of trace/ultratrace rare earth elements in environmental samples by ICP MS after MSPE with $Fe_3O_4@SiO_2$@polyaniline–graphene oxide composite. Talanta 119:458–466
70. Wu S, Hu C, He M, Chen B, Hu B (2013) Capillary microextraction combined with fluorinating assisted ETV ICP-OES for the determination of trace La, Eu, Dy and Y in human hair. Talanta 115:342–348
71. Pyrzynska K (2010) Carbon nanostructures for separation, preconcentration and speciation of metal ions. Trends Anal Chem 29:718–727
72. Rybak A, Rybak A, Kaszuwara W, Boncel S (2018) Poly (2,6-dimethyl-1,4-phenylene oxide) hybrid membranes filled with magnetically aligned iron-encapsulated carbon nanotubes (Fe@MWCNTs) for enhanced air separation. Diam Relat Mater 83:21–29
73. Tajabadi F, Yamini Y, Sovizi MR (2013) Carbon-based magnetic nanocomposites in solid phase dispersion for some of lanthanides, followed by their quantitation via ICP-OES. Microchim Acta 180:65–73

74. Rybak A, Rybak A, Boncel S, Kolanowska A, Jakobik-Kolon A, Bok-Badura J (2024) The modern rare earth imprinted membranes for REE metal ions recovery from coal fly ashes extracts. Materials 17:3087
75. Rybak A, Rybak A, Kolanowska A (2024) Jonowo imprintowane membrany hybrydowe z dodatkiem funkcjonalizowanych MWCNT i ich zastosowanie do odzysku jonów REE z ekstraktów popiołów lotnych. In: Interdyscyplinarność kluczem do rozwoju. Fundacja na rzecz promocji nauki i rozwoju TYGIEL, Lublin. ISBN 978-83-67670-53-1
76. Rybak A, Rybak A, Kolev SD (2023) A modern computer application to model rare earth element ion behavior in adsorptive membranes and materials. Membranes 13:175–187
77. Rybak A, Rybak A, Boncel S (2024) Nowoczesne aplikacje komputerowe REE_isotherm i REE 2.0 i ich zastosowanie do analizy zachowania jonów REE w wybranych materiałach adsorpcyjnych. In: Interdyscyplinarność kluczem do rozwoju. Fundacja na rzecz promocji nauki i rozwoju TYGIEL, Lublin. ISBN 978-83-67670-53-1
78. Dolak I, Kecili R, Hur D, Ersoz A, Say R (2015) Ion-imprinted polymers for selective recognition of neodymium (III) in environmental samples. Ind Eng Chem Res 54:5328–5335
79. Zheng X, Zhang Y, Bian T et al (2019) One-step fabrication of imprinted mesoporous cellulose nanocrystals films for selective separation and recovery of Nd (III). Cellulose 26:5571–5582
80. Krishna PG, Gladis JM, Rao TP, Naidu GR (2005) Selective recognition of neodymium (III) using ion imprinted polymer particles. J Mol Recognit 18:109–116
81. Krishna X, Zhang F, Liu E, Xu X, Yan Y (2017) Efficient recovery of neodymium in acidic system by free-standing dual template docking oriented ionic imprinted mesoporous films. ACS Appl Mater Interfaces 9:730–739
82. Melnyk IV, Goncharyk VP, Kozhara LI, Yurchenko GR, Matkovsky AK, Zub YL, Alonso B (2012) Sorption properties of porous spray-dried microspheres functionalized by phosphonic acid groups. Microporous Mesoporous Mater 153:171–177
83. Galhoum AA, Atia AA, Mahfouz MG (2015) Dy (III) recovery from dilute solutions using magnetic-chitosan nano-based particles grafted with amino acids. J Mater Sci 50:2832–2848
84. Zheng X, Liu E, Zhang F (2016) Efficient adsorption and separation of dysprosium from NdFeB magnets in an acidic system by ion imprinted mesoporous silica sealed in a dialysis bag. Green Chem 18:5031–5040
85. Galhoum AA, Mahfouz MG, Abdel-Rehem ST (2015) Diethylenetriamine-functionalized chitosan magnetic nano-based particles for the sorption of rare earth metal ions [Nd (III), Dy (III) and Yb (III)]. Cellulose 22:2589–2605
86. Shojaei Z, Iravani E, Moosavian MA, Torab-Mostaedi M (2016) Removal of cerium from aqueous solutions by amino phosphate modified nano TiO_2: kinetic, and equilibrium studies. Iran J Chem Eng 13(2):3
87. Keçili R, Dolak I, Ziyadanoguları B, Ersoz A, Say R (2018) Ion imprinted cryogel-based super macroporous traps for selective separation of cerium (III) in real samples. J Rare Earths 36:857–862
88. Liu J, Yang XL, Cheng XZ, Peng Y, Chen HM (2013) Synthesis and application of ion-imprinted polymer particles for solid-phase extraction and determination of trace scandium by ICP-MS in different matrices. Anal Methods 5:1811–1817
89. Gao BJ, Zhang YQ, Xu Y (2014) Study on recognition and separation of rare earth ions at picometre scale by using efficient ion-surface imprinted polymer materials. Hydrometallurgy 150:83–91
90. Vigneau O, Pinel C, Lemaire M (2001) Ionic imprinted resins based on EDTA and DTPA derivatives for lanthanides (III). Anal Chim Acta 435:75–82

Chapter 7
Summary

Abstract The presented book addresses issues related to clean coal technologies and their role in energy transformation. The analyses carried out concern the most important countries still based on coal, i.e. China, India, Australia, USA, with particular emphasis on the European Union. The study focuses on coal, because clean coal technologies were created for coal and it is mainly coal that is the subject of most global decarbonization treaties. Poland, as one of the EU member states, has an exceptionally complicated task - energy transformation and decarbonization of the energy system by 2050. The restrictive standards and expectations of the EU towards its members put Poland in an unprecedented position on a global scale. Therefore, Poland was considered the most representative country in this case. Either Poland will apply CCT and renewable energy and will turn out to be a great winner, or it will fail during the energy transformation process, the main consequences of which will fall on its citizens. The book presents the most promising CCT solutions, future directions of CCT development, the role of CCT in energy transformation and the development of renewable energy sources. If the implementation of CCT is successful in Poland, there is a high probability that their introduction will be effective in every country in the world.

Keywords Clean coal technologies · Energy transformation · Energy clusters · REE · Renewable energy sources

Clean coal technologies are technologically mature solutions that can be used in the period of the global energy transition. Work on their development and improvement is still ongoing. More and more efficient solutions are being developed that eliminate the shortcomings of previous solutions. The authors of the presented study are working on membrane techniques that are useful in the process of capturing and recovering gaseous and solid waste from the coal combustion process. CCT contributes greatly to the environmental goals set in the Paris Agreement and related documents. According to the UN, clean coal technologies can support the implementation of sustainable development goals. This study presents several possible technological and organisational solutions. As shown, these solutions, if implemented, would make it possible to support the energy transition period with fossil fuels in an ecological, economical, and justified manner from the point of view of energy security. CCT would provide excellent support for the development of renewable energy sources by providing the raw materials necessary for the construction of their infrastructure, the construction of energy storage facilities, stabilising the operation of the energy system, and means of financing transformation. The future of CCT will depend to the greatest extent on energy policy, at the global level, which shapes the policies of individual countries. Energy policy will determine whether CCT will get their chance and we will go through the period of energy transition using CCT, gradually increasing the share of RES, or whether it will be a sudden and short-term process. This study presents a simulation that illustrates the potential effects of a radical and sudden departure from coal fuel. It has been proven that such an action would have a disastrous impact on the level of energy security, which is of key importance to states and societies in times of social unrest and wars. The main problem of energy transition in many countries is the lack of time resulting from many years of neglect on the path to decarbonization. Clean coal technologies would allow to gain time to develop RES solutions that could function in the future as fully independent, stable, safe energy sources that eliminate the current weaknesses of energy systems based on RES.